Image Analysis
of
Food
Microstructure

Image Analysis
of
Food
Microstructure

John C. Russ

CRC Press
Taylor & Francis Group
Boca Raton London New York

CRC Press is an imprint of the
Taylor & Francis Group, an **informa** business

CRC Press
Taylor & Francis Group
6000 Broken Sound Parkway NW, Suite 300
Boca Raton, FL 33487-2742

First issued in paperback 2019

ISBN-13: 978-0-8493-2241-9 (hbk)
ISBN-13: 978-0-367-39359-5 (pbk)

Library of Congress Cataloging-in-Publication Data

Russ, John C.
  Image analysis of food microstructure / John C. Russ.
    p. cm.
  Includes index.
  ISBN 0-8493-2241-3 (alk. paper)
    1. Food—Analysis. 2. Microscopy. 3. Image analysis. I. Title.

  TX543.R88 2004
  664'.07—dc22

                                                                2004051958

Library of Congress Card Number 2004051958

Visit the Informa Web site at
www.informa.com

and the Informa Healthcare Web site at
www.informahealthcare.com

Contents

Chapter 1 Stereology.. 1

The Need for Stereology 1
Unfolding Size Distributions... 5
Volume Fraction.. 11
Surface Area.. 21
Lines and Points 28
Design of Experiments 31
Topological Properties 34
Other Stereological Techniques.. 41

Chapter 2 Image Acquisition 51

Scanners... 51
Digital Cameras 57
Scanning Microscopes 63
File Formats... 67
Color Adjustment.. 71
Color Space Coordinates 77
Color Channels 81
Optimum Image Contrast 86
Removing Noise 98
Nonuniform Illumination.. 111
Image Distortion and Focus 121
Summary.. 127

Chapter 3 Image Enhancement 129

Improving Local Contrast... 131
Image Sharpening 135
False Color and Surface Rendering 142
Rank-Based Filters 145
Edge-Finding... 147
Texture 156
Directionality 164
Finding Features in Images 169
Image Combinations.. 175
Thresholding 184
Automatic Threshold Settings Using the Histogram 189
Automatic Thresholding Using the Image 197

Other Thresholding Approaches..200
Color Image Thresholding...205
Manual Marking 209
Summary 211

Chapter 4   Binary Images 213

Erosion and Dilation..213
The Euclidean Distance Map 219
Separating Touching Features 223
Boolean Combinations 236
Using Grids for Measurement...240
Using Markers to Select Features 247
Combined Boolean Operations 250
Region Outlines as Selection Criteria...252
Skeletonization..254
Fiber Images 260
Skeletons and Feature Shape...262
Measuring Distances and Locations with the EDM 263
Summary 275

Chapter 5   Measuring Features 277

Counting...277
Calibration...287
Size Measurement..292
Size Distributions..297
Comparisons 301
Edge Correction...303
Brightness and Color Measurements 310
Location 316
Gradients..324
Shape..342
Identification 350
Conclusions..359

Index..361

Introduction

Why is a book about food microstructure written by someone with a background in engineering materials?

My own path to recognizing the importance of image analysis to measure microstructural parameters in food products has been round-about. For nearly two decades as a professor in the Materials Science and Engineering Department at North Carolina State University I taught a sophomore course in basic materials science that included such fundamentals as mechanical properties of metals, ceramics, polymers and composites and the procedures for testing them, the basics of phase diagrams and the formation of microstructures during phase transformations, and so on. Because many of the students came into the course with little background in materials, but with about 20 years of experience in eating, I commonly used food examples to explain various processes. College students generally respond well to comparisons of using the liquid-solid phase change to produce ice beer versus the liquid-gas phase change to distill corn liquor. Other examples used cooked spaghetti as a model for the entanglement of linear polymers, and fruit in Jell-O as a model for dispersion strengthening.

Little did I realize at the time that the people whose products are food would look to materials for models of behavior. It was only recently that I discovered in the excellent text by Aguilera and Stanley (*Microstructural Principles of Food Processing and Engineering, 2nd edition,* Aspen Press, 1999) the identical descriptions and equations from materials science as relating to food products. The same basic understanding pervades both fields, namely that product performance (mechanical, chemical, environmental, nutritive, etc.) depends on structure, and that structure in turn depends on processing history. For most engineering materials this is limited to chemical composition and thermo-mechanical processing, but for food products it also includes genetics — breeding plants and animals to produce desirable structural properties.

In all cases, there is great interest in quantifying the various aspects of structure, and in many cases this involves imaging the structure and making measurements on the images. The images may be simple macroscopic or microscopic light images, including confocal light microscopy, but can also include electron images (with either the transmission or scanning electron microscope — SEM and TEM), atomic force microscope (AFM) images of surfaces, magnetic resonance (MRI) or computed tomography (CT) images of internal structure, and many more. Some techniques that we do not normally think of as imaging produce data sets that may be best interpreted as images, by plotting the data so that the eye (and the computer, with appropriate software) can identify trends, optima, etc. Even a simple one-dimensional graph such as a stress-strain curve may reveal more (e.g., the fractal irregularities of the curve) to the eye than the column of numbers from which it was plotted, and this becomes more significant for two-dimensional pictures (but unfortunately becomes more difficult for higher dimensionalities, because of the problems of plotting and viewing n-dimensional data).

In addition to teaching undergraduates about the basics of materials, I also taught a graduate course in image analysis and stereology. Although listed as a materials science course, this consistently attracted significant numbers of students from other majors, including textiles, wood and paper products, biology, the vet school, even archaeology — and a steady trickle of food science students. So eventually I was asked to sit on several graduate student advisory committees in the Food Science Department, got to know some of the faculty there, suggested various ways that image analysis could be used to measure structures of interest, and in the process got something of an education about food microstructure (and became more aware — not necessarily in a positive way — of what I was personally consuming).

That in turn led to contacts with other researchers, at other universities such as Penn State and the University of Guelph, to societies and organizations such as the Food Structure and Functionality Forum (a division of the AOCS), Agriculture and Agri-Food Canada, and the Hydrocolloids conferences, and to various corporations and their food interests, ranging from tiny (a specialty chocolate manufacturer and the producer of a nutraceutical supplement with microencapsulated omega-3 fatty acids) to large (major producers of cake mixes, baked goods, processed meats, and so on). During this process, I have been continually surprised to see the same questions and problems surfacing over and over. While the food products themselves, and to some extent the images and terminology, differ widely, the structural properties of interest, and the appropriate ways to determine them from images, tend to be much the same.

That gives me hope that this book can usefully summarize the basic procedures that will be useful to many of these researchers. The topics covered are the acquisition and processing of the images, the measurement of appropriate microstructural parameters, and the interpretation of those numbers required by the fact that the structures are generally three-dimensional while the images are usually two-dimensional. In most general textbooks in image analysis, including my own (*The Image Processing Handbook, 4th edition*, CRC Press, 2002), the organization typically begins with the characteristics of cameras, proceeds through the various processing steps on color or grey scale images, and then discusses segmentation or thresholding and the processing and measurement of binary images. The data from these measurements is then used as the subject for statistical analysis and perhaps the construction of expert systems for feature recognition.

As a framework for instruction, that sequential organization is useful, but for this text I am risking a different approach. This book starts with basic stereology, which is the essentially geometric science that relates three-dimensional structures to the measurements that can be made on two-dimensional slices such as typical microscope images. From this consideration emerges the fundamental ideas of what *can* and *should* be measured on structures, and that will guide subsequent chapters of the book. It is conventional to think about measurement as the last step, after processing and thresholding. But in complex structures it is often very useful to obtain measurement data directly from the processing operations themselves. Consequently, measurements will be introduced throughout the various chapters. The reader is invited to relate these measurements to the important history-structure-function relationships in the particular kinds of food products of personal interest.

Also, processing of images in the "pixel" or spatial domain is often considered separately from processing in the frequency or Fourier domain, because the math appears rather different. Since the math is largely suppressed in this book anyway (which emphasizes the underlying concepts and concentrates on the visual interpretation of the results), I have decided to avoid that separation and mix both approaches together. The resulting organization is based on what I have found to be a useful step-by-step approach to extraction of information from images, largely driven by what information is being sought.

This book does not contain the usual hundreds of literature citations for the various procedures described. The presentation here does not pretend to include the technical details of the various algorithms and procedures, just illustrations of how they are used. There are plenty of texts, my own included, that do have those references, as well as fuller explanations of the underlying math and programs. For anyone planning to write their own programs, reference to those texts will be necessary anyway and will lead directly to the primary sources. The intent here is to familiarize the food scientist with ways of thinking about images, their processing and measurement. To borrow a description one of my students once used, it is about playing music, not writing it.

Also, there is no list of citations to work that has used these methods to measure food structures. My rationale for this absence is that I am involved with the measurement process and not the food science, and am reluctant to judge which papers are meaningful and which are not. It is my impression that there has been some very good work done from a measurement standpoint (as well as much that is questionable), but whether that translates to a better understanding of the processing and properties of the food structures is beyond my range of understanding, in spite of the best efforts of some folks acknowledged below to educate me.

Finally, a word of caution to the reader: This is not something you can really learn by reading through the book and thinking about it — you need to do it yourself. To continue the analogy from above, you have to play the music, not just listen to it. Learning from words and pictures in a book is no substitute for learning by trying out the methods on real images and seeing the results of step-by-step procedures. It helps to have your own images of the structures of particular interest. And, of course, that means you need the appropriate cameras, computers, software and so on. Most of the techniques described in the book can be performed using a wide variety of commercially available software, ranging from expensive dedicated image processing and analysis packages such as Media Cybernetics' Image Pro Plus or Universal Imaging's Metamorph programs, to data handling environments such as Matlab which can also handle images.

That is not intended as an endorsement and is certainly not a comprehensive list — there are literally hundreds of companies who sell image processing or analysis software, some highly specialized for particular niche markets. Each program will impose some limitations on what you can accomplish but more importantly will typically use quite different terminology to describe the operations, and will have the functions organized in entirely different ways. Learning the basics of what you can do and want to do from this book is only the first step; then you have to study the software manual to find out how to do it.

As an aid to the researcher who does not already have software in place, or who finds that their particular package lacks some of the tools described here, I have collaborated with Reindeer Graphics, Inc. (http://www.ReindeerGraphics.com), a company run by my son and by one of my former students, to produce a set of Photoshop®-compatible plugins that implement each of the algorithms used in the text. Information on the software is available online; the CD that installs the software includes a lengthy tutorial showing how to use the various procedures, with a large suite of test images for instructional purposes (the software can also be used with any other images). There are two packages: Fovea Pro and The Image Processing Tool Kit. The Tool Kit is intended as a low-cost educational package, limited to 8 bit per channel grey scale and 24 bit RGB color images, adequate for bright field microscopy and most SEM images. Fovea Pro includes all of the Tool Kit functions and also works with 16 bit per channel images (important for dark field and fluorescence images, transmission EM pictures, and most surface imaging data) including 48 bit RGB color (such as the output from most film or flatbed scanners).

Isolating each function as a separate menu item facilitates learning what each method does, but the software is not limited to teaching. It can also be used for real analysis, and sequences of operations can easily be created (e.g., using Photoshop® Actions) to carry out complicated automatic analysis of batches of images. Adobe Photoshop was selected as a platform for this because it is relatively inexpensive, well documented and with a reasonable learning curve, and already in widespread use in labs for image acquisition, presentation, annotation, etc. Photoshop supports a wide variety of image formats and acquisition devices (scanners, cameras, etc.). Also, the fact that Photoshop is so widely used means that other programs have adopted the same convention for add-on plugins, and Photoshop-compatible plugins can also be used by many other programs on both Windows and Macintosh computers, ranging from inexpensive software such as Jasc's Paint Shop Pro to costly professional software such as Media Cybernetics' Image Pro Plus.

My special thanks go out to the many food scientists who have contributed generously to this book. Their support has ranged from general encouragement for the idea that it was time that such a text was available, to taking the time to educate me about a particular food product or problem, to providing example images. I can not begin to list here everyone who has helped, but I have tried to credit each picture supplied by someone else in the captions (and I apologize if I have missed someone unintentionally due to my own poor record keeping!). Special thanks are due to Allen Foegeding at North Carolina State University for getting this whole project started, reviewing the text to try to keep me from saying nonsensical things about food, and helping a lot along the way; to José Aguilera and David Stanley for writing their book and encouraging mine; to Alex Marangoni, Howard Swatland and others at the University of Guelph for sharing their library of many years of images; to Ken Baker for providing access to his image collection; and to Greg Ziegler at Penn State University and Paula Allan-Wojtas at Agriculture and Agri-Food Canada, for images, ideas, and invitations.

John Russ
Raleigh, NC

1 Stereology

THE NEED FOR STEREOLOGY

Before starting with the process of acquiring, correcting and measuring images, it seems important to spend a chapter addressing the important question of just what it is that can and should be measured, and what cannot or should not be. The temptation to just measure everything that software can report, and hope that a good statistics program can extract some meaningful parameters, is both naïve and dangerous. No statistics program can correct, for instance, for the unknown but potentially large bias that results from an inappropriate sampling procedure.

Most of the problems with image measurement arise because of the nature of the sample, even if the image itself captures the details present perfectly. Some aspects of sampling, while vitally important, will not be discussed here. The need to obtain a representative, uniform, randomized sample of the population of things to be measured should be obvious, although it may be overlooked, or a procedure used that does not guarantee an unbiased result. A procedure, described below, known as systematic random sampling is the most efficient way to accomplish this goal once all of the contributing factors in the measurement procedure have been identified.

In some cases the images we acquire are of 3D objects, such as a dispersion of starch granules or rice grains for size measurement. These pictures may be taken with a macro camera or an SEM, depending on the magnification required, and provided that some care is taken in dispersing the particles on a contrasting surface so that small particles do not hide behind large ones, there should be no difficulty in interpreting the results. Bias in assessing size and shape can be introduced if the particles lie down on the surface due to gravity or electrostatic effects, but often this is useful (for example, measuring the length of the rice grains).

Much of the interest in food structure has to do with internal microstructure, and that is typically revealed by a sectioning procedure. In rare instances volume imaging is performed, for instance, with MRI or CT (magnetic resonance imaging and computerized tomography), both techniques borrowed from medical imaging. However, the cost of such procedures and the difficulty in analyzing the resulting data sets limits their usefulness. Full three-dimensional image sets are also obtained from either optical or serial sectioning of specimens. The rapid spread of confocal light microscopes in particular has facilitated capturing such sets of data. For a variety of reasons — resolution that varies with position and direction, the large size of the data files, and the fact that most 3D software is more concerned with rendering visual displays of the structure than with measurement — these volume imaging results are not commonly used for structural measurement.

Most of the microstructural parameters that robustly describe 3D structure are more efficiently determined using stereological rules with measurements performed

1

on section images. These may be captured from transmission light or electron microscopes using thin sections, or from light microscopes using reflected light, scanning electron microscopes, and atomic force microscopes (among others) using planar surfaces through the structure. For measurements on these images to correctly represent the 3D structure, we must meet several criteria. One is that the surfaces are properly representative of the structure, which is sometimes a non-trivial issue and is discussed below. Another is that the relationships between two and three dimensions are understood.

That is where stereology (literally the study of three dimensions, and unrelated to stereoscopy which is the viewing of three dimensions using two eye views) comes in. It is a mathematical science developed over the past four decades but with roots going back two centuries. Deriving the relationships of geometric probability is a specialized field occupied by a few mathematicians, but using them is typically very simple, with no threatening math. The hard part is to understand and visualize the meaning of the relationships and recognizing the need to use them, because they tell us what to measure and how to do it. The rules work at all scales from nm to light-years and are applied in many diverse fields, ranging from materials science to astronomy.

Consider for example a box containing fruit — melons, grapefruit and plums — as shown in Figure 1.1. If a section is cut through the box and intersects the fruit, then an image of that section plane will show circles of various colors (green, yellow and purple, respectively) that identify the individual pieces of fruit. But the sizes of the circles are not the sizes of the fruit. Few of the cuts will pass through the equator of a spherical fruit to produce a circle whose diameter would give the size of the sphere. Most of the cuts will be smaller, and some may be very small where the plane of the cut is near the north or south pole of the sphere. So measuring the 3D sizes of the fruit is not possible directly.

What about the number of fruits? Since they have unique colors, does counting the number of intersections reveal the relative abundance of each type? No. Any plane cut through the box is much more likely to hit a large melon than a small plum. The smaller fruits are under-represented on the plane. In fact, the probability of intersecting a fruit is directly proportional to the diameter. So just counting doesn't give the desired information, either.

Counting the features present can be useful, if we have some independent way to determine the mean size of the spheres. For example, if we've already measured the sizes of melons, plums and grapefruit, then the number per unit volume N_V of each type fruit in the box is related to the number of intersections per unit area N_A seen on the 2D image by the relationship

$$N_V = \frac{N_A}{D_{mean}} \tag{1.1}$$

where D_{mean} is the mean diameter.

In stereology the capital letter N is used for number and the subscript V for volume and A for area, so this would be read as "Number per unit volume equals

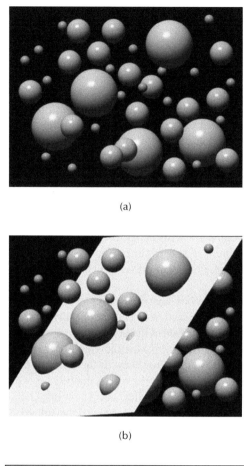

(a)

(b)

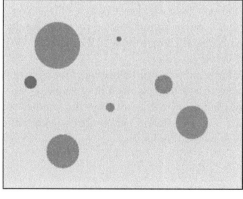

(c)

FIGURE 1.1 (See color insert following page 150.) Schematic diagram of a box containing fruit: (a) green melons, yellow grapefruit, purple plums; (b) an arbitrary section plane through the box and its contents; (c) the image of that section plane showing intersections with the fruit.

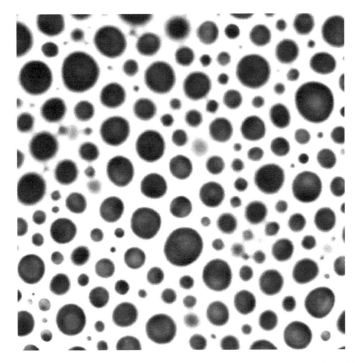

FIGURE 1.2 Section image through a foamed food product. (Courtesy of Allen Foegeding, North Carolina State University, Department of Food Science)

number per unit area divided by mean diameter." Rather than using the word "equals" it would be better to say "is estimated by" because most of the stereological relationships are statistical in nature and the measurement procedure and calculation give a result that (like all measurement procedures) give an estimate of the true result, and usually a way to also determine the precision of the estimate.

The formal relationship shown in Equation 1.1 relates the expected value (the average of many observed results) of the number of features per unit area to the actual number per unit volume times the mean diameter. For a series of observations (examination of multiple fields of view) the average result will approach the expected value, subject to the need for examining a representative set of samples while avoiding any bias. Most of the stereological relationships that will be shown are for expected values.

Consider a sample like the thick-walled foam in Figure 1.2 (a section through a foamed food product). The size of the bubbles is determined by the gas pressure, liquid viscosity, and the size of the hole in the nozzle of the spray can. If this mean diameter is known, then the number of bubbles per cubic centimeter can be calculated from the number of features per unit area using Equation 1.1. The two obvious things to do on an image like those in Figures 1.1 and 1.2 are to count features and measure the sizes of the circles, but both require stereological interpretation to yield a meaningful result.

This problem was recognized long ago, and solutions have been proposed since the 1920s. The basic approach to recovering the size distribution of 3D features from the image of 2D intersections is called "unfolding." It is now out of favor with most stereologists because of two important problems, discussed below, but since it is still useful in some situations (and is still used in more applications than it probably should be), and because it illustrates an important way of thinking about three dimensions, a few paragraphs will be devoted to it.

UNFOLDING SIZE DISTRIBUTIONS

Random intersections through a sphere of known radius produce a distribution of circle sizes that can be calculated analytically as shown in Figure 1.3. If a large number of section images are measured, and a size distribution of the observed circles is determined, then the very largest circles can only have come from near-equatorial cuts through the largest spheres. So the size of the largest spheres is established, and their number can be calculated using Equation 1.1.

But if this number of large spheres is present, the expected number of cross sections of various different smaller diameters can be calculated using the derived relationship, and the corresponding number of circles subtracted from each smaller bin in the measured size distribution. If that process leaves a number of circles remaining in the next smallest size bin, it can be assumed that they must represent near-equatorial cuts through spheres of that size, and their number can be calculated. This procedure can be repeated for each of the smaller size categories, typically 10 to 15 size classes. Note that this does not allow any inference about the size sphere that corresponds to any particular circle, but is a statistical relationship that depends upon the collective result from a large number of intersections.

If performed in this way, a minor problem arises. Because of counting statistics, the number of circles in each size class has a finite precision. Subtracting one number

FIGURE 1.3 Schematic diagram of sectioning a sphere to produce circles of different sizes.

(the expected number of circles based on the result in a larger class) from another (the number of circles observed in the current size class) leaves a much smaller net result, but with a much larger statistical uncertainty. The result of the stepwise approach leads to very large statistical errors accumulating for the smallest size classes.

That problem is easily solved by using a set of simultaneous equations and solving for all of the bins in the distribution at the same time. Tables of coefficients that calculate the number of spheres in each size class (i) from the number of circles in size class (j) have been published many times, with some difference depending on how the bin classes are set up. One widely used version is shown in Table 1.1. The mathematics of the calculation is very simple and easily implemented in a spreadsheet. The number of spheres in size class i is calculated as the sum of the number of circles in each size class j times an alpha coefficient (Equation 1.2). Note that half of the matrix of alpha values is empty because no large circles can be produced by small spheres.

$$N_{V_i} = \sum_j \alpha_{ij} \cdot N_{A_j} \qquad (1.2)$$

Figure 1.4 shows the application of this technique to the bubbles in the image of Figure 1.2. The circle size distribution shows a wide variation in the sizes of the intersections of the bubbles with the section plane, but the calculated sphere size distribution shows that the bubbles are actually all of the same size, within counting statistics. Notice that this calculation does not directly depend on the actual sizes of the features, but just requires that the size classes represent equal-sized linear increments starting from zero.

Even with the matrix solution of all equations at the same time, this is still an ill conditioned problem mathematically. That means that because of the subtractions (note that most of the alpha coefficients are negative, carrying out the removal of smaller circles expected to correspond to larger spheres) the statistical precision of the resulting distribution of sphere sizes is much larger (worse) than the counting precision of the distribution of circle sizes. Many stereological relationships can be estimated satisfactorily from only a few images and a small number of counts. However, unfolding a size distribution does not fit into this category and very large numbers of raw measurements are required.

The more important problem, which has led to the attempts to find other techniques for determining 3D feature sizes, is that of shape. The alpha matrix values depend critically on the assumption that the features are all spheres. If they are not, the distribution of sizes of random intersections changes dramatically. As a simple example, cubic particles produce a very large number of small intersections (where a corner is cut) and the most probable size is close to the area of a face of the cube, not the maximum value that occurs when the cube is cut diagonally (a rare event). For the sphere, on the other hand, the most probable value is large, close to the equatorial diameter, and very small cuts that nip the poles of the sphere are rare, as shown in Figure 1.5.

TABLE 1.1
Matrix of Alpha Values Used to Convert the Distribution of Number of Circles per Unit Area to Number of Spheres per Unit Volume

$N_A(1)$	$N_A(2)$	$N_A(3)$	$N_A(4)$	$N_A(5)$	$N_A(6)$	$N_A(7)$	$N_A(8)$	$N_A(9)$	$N_A(10)$	$N_A(11)$	$N_A(12)$	$N_A(13)$	$N_A(14)$	$N_A(15)$
0.26491	−0.19269	0.01015	−0.01636	−0.00538	−0.00481	−0.00327	−0.00250	−0.00189	−0.00145	−0.00109	−0.00080	−0.00055	−0.00033	−0.00013
	0.27472	−0.19973	0.01067	−0.01691	−0.00549	−0.00491	−0.00330	−0.00250	−0.00186	−0.00139	−0.00101	−0.00069	−0.00040	−0.00016
		0.28571	−0.20761	0.01128	−0.01751	−0.00560	−0.00501	−0.00332	−0.00248	−0.00180	−0.0012	−0.00087	−0.00051	−0.00020
			0.29814	−0.21649	0.01200	−0.01818	−0.00571	−0.00509	−0.00332	−0.00242	−0.00169	−0.00113	−0.00066	−0.00026
				0.31235	−0.22663	0.01287	−0.01893	−0.00579	−0.00516	−0.00327	−0.00230	−0.00150	−0.00087	−0.00034
					0.32880	−0.23834	0.01393	−0.01977	−0.00584	−0.00518	−0.00315	−0.00208	−0.00117	−0.00045
						0.34816	−0.25208	0.01527	−0.02071	−0.00582	−0.00512	−0.00288	−0.00167	−0.00062
							0.37139	−0.26850	0.01704	−0.02176	−0.00565	−0.00488	−0.00234	−0.00094
								0.40000	−0.28863	0.01947	−0.02293	−0.00516	−0.00427	−0.00126
									0.43644	−0.31409	0.02308	−0.02416	−0.00393	−0.00298
										0.48507	−0.34778	0.02903	−0.02528	−0.00048
											0.55470	−0.39550	0.04087	−0.02799
												0.66667	−0.47183	0.08217
													0.89443	−0.68328
														1.00000

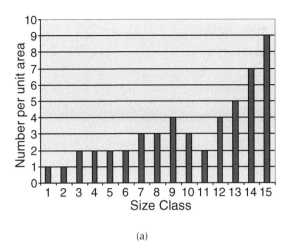

(a)

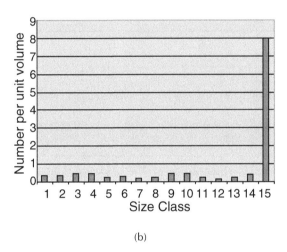

(b)

FIGURE 1.4 Calculation of sphere sizes: (a) measured circle size distribution from Figure 1. 2; (b) distribution of sphere sizes calculated from a using Equation 1.2 and Table 1.1. The plots show the relative number of objects as a function of size class.

In theory it is possible to compute an alpha matrix for any shape, and copious tables have been published for a wide variety of polygonal, cylindrical, ellipsoidal, and other geometric shapes. But the assumption still applies that all of the 3D features present have the same shape, and that it is known. Unfortunately, in real systems this is rarely the case (see the example of the pores, or "cells" in the bread in Figure 1.6). It is very common to find that shapes vary a great deal, and often vary systematically with size. Such variations invalidate the fundamental approach of size unfolding.

That the unfolding technique is still in use is due primarily to two factors: first, there really are some systems in which a sphere is a reasonable model for feature shape. These include liquid drops, for instance in an emulsion, in which surface tension produces a spherical shape. Figure 1.7 shows spherical fat droplets in

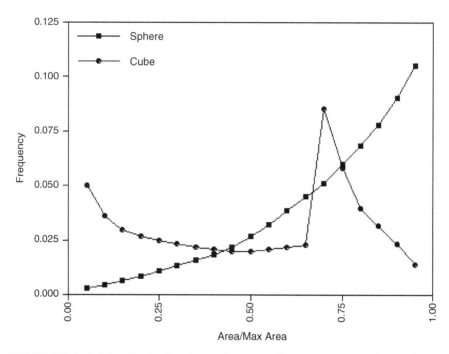

FIGURE 1.5 Probability distributions for sections through a sphere compared to a cube.

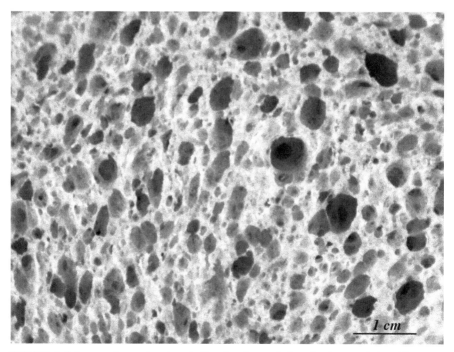

FIGURE 1.6 Image of pores in a bread slice showing variations in shape and size. (Courtesy of Diana Kittleson, General Mills)

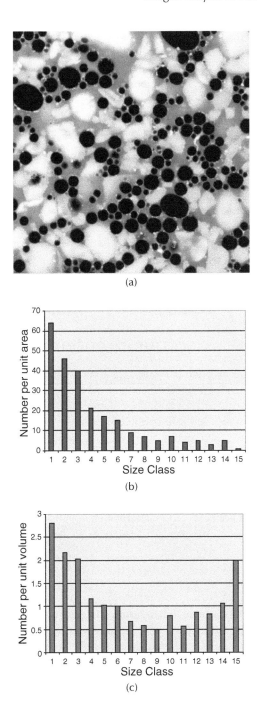

(a)

(b)

(c)

FIGURE 1.7 Calculation of sphere size distribution: (a) image of fat droplets in mayonnaise (Courtesy of Anke Janssen, ATO B.V., Food Structure and Technology); (b) measured histogram of circle sizes; (c) calculated distribution of sphere sizes. The plots show the relative number of objects as a function of size class.

mayonnaise, for which the circle size distribution can be processed to yield a distribution of sphere sizes. Note that some of the steps needed to isolate the circles for measurement will be described in detail in later chapters.

The second reason for the continued use of sphere unfolding is ignorance, laziness and blind faith. The notion that "maybe the shapes aren't really spheres, but surely I can still get a result that will compare product A to product B" is utterly wrong (different shapes are likely to bias the results in quite unexpected ways). But until researchers gain familiarity with some of the newer techniques that permit unbiased measurement of the size of three-dimensional objects they are reluctant to abandon the older method, even if deep-down they know it is not right.

Fortunately there are methods, such as the point-sampled intercept and disector techniques described below, that allow the unbiased determination of three-dimensional sizes regardless of shape. Many of these methods are part of the so-called "new stereology," "design-based stereology," or "second-order stereology" that have been developed within the past two decades and are now becoming more widely known. First, however, it will be useful to visit some of the "old" stereology, classical techniques that provide some very important measures of three-dimensional structure.

VOLUME FRACTION

Going back to the structure in Figure 1.2, if the sphere size is known, the number can be calculated from the volume fraction of bubbles, which can also be measured from the 2D image. In fact, determining volume fraction is one of the most basic stereological procedures, and one of the oldest. A French geologist interested in determining the volume fraction of ore in rock 150 years ago, realized that the area fraction of a section image that showed the ore gave the desired result. The stereologists' notation represents this as Equation 1.3, in which the V_V represents the volume of the phase or structure of interest per unit volume of sample, and the A_A represents the area of that phase or structure that is visible in the area of the image. As noted before, this is an expected value relationship that actually says the expected value of the area fraction observed will converge to the volume fraction.

$$V_V = A_A \tag{1.3}$$

To understand this simple relationship, imagine the section plane sweeping through a volume; the area of the intersections with the ore integrates to the total volume of ore, and the area fraction integrates to the volume fraction. So subject to the usual caveats about requiring representative, unbiased sampling, the expected value of the area fraction is (or measures) the volume fraction.

In the middle of the nineteenth century, the area fraction was not determined with digital cameras and computers, of course; not even with traditional photography, which had only just been invented and was not yet commonly performed with microscopes. Instead, the image was projected onto a piece of paper, the features of interest carefully traced, and then the paper cut and weighed. The equivalent modern measurement of area fraction can often be accomplished by counting pixels in the image histogram, as shown in Figure 1.8. The histogram is simply a plot of the number of pixels having each of the various brightness levels in the image, often 256. The interpretation of the histogram will be described in subsequent chapters. Although very efficient, this is not always the preferred method for measurement of volume fraction, because the precision of the measurement is better estimated using other approaches.

The measurement of the area represented by peaks in the histogram is further complicated by the fact that not all of the pixels in the image have brightness values that place them in the peaks. As shown in Figure 1.9, there is generally a background level between the peaks that can represent a significant percentage of the total image area. In part this is due to the finite area of each pixel, which averages the information from a small square on the image. Also, there is usually some variation in pixel brightness (referred to generally as noise) even from a perfectly uniform area. Chapter 3 discusses techniques for reducing this noise. Notice that this image is not a photograph of a section, but has been produced non-destructively by X-ray tomography. The brightness is a measure of local density.

The next evolution in methodology for measuring volume fraction came fifty years after the area fraction technique, again introduced as a way to measure minerals. Instead of measuring areas, which is difficult, a random line was drawn on the image and the length of that line which passed through the structure of interest was measured (Figure 1.10). The line length fraction is also an estimate of the volume fraction. For understanding, imagine the line sweeping across the image; the line length fraction integrates to the area fraction. The stereological notation is shown in Equation 1.4, where L_L represents the length of the intersections as a fraction of the total line length.

$$V_V = L_L \qquad\qquad (1.4)$$

The advantage of this method lies in the greater ease with which the line length can be measured as compared to area measurements. Even in the 1950s my initial experience with measurement of volume fraction used this approach. A small motor was used to drive the horizontal position of a microscope stage, with a counter keeping track of the total distance traveled. Another counter could be engaged by pressing a key, which the human observer did whenever the structure of interest was passing underneath the microscope's crosshairs. The ratio of the two counter numbers gave the line length fraction, and hence the volume fraction. Replacing the human eye with an electronic sensor whose output could be measured to identify the phase created an automatic image analyzer.

By the middle of the twentieth century, a Russian metallurgist had developed an even simpler method for determining volume fraction that avoided the need to make a measurement of area or length, and instead used a counting procedure. Placing a grid of points on the image of the specimen (Figure 1.11), and counting the fraction of them that fall onto the structure of interest, gives the point fraction

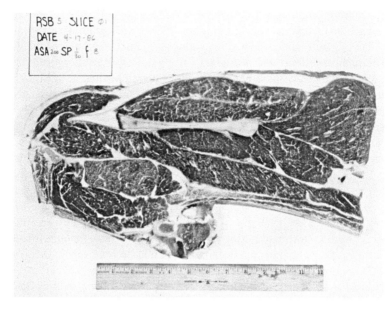

(a)

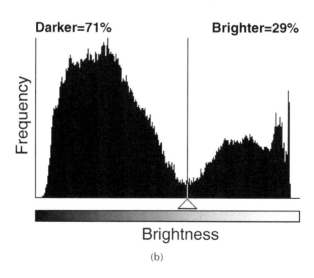

(b)

FIGURE 1.8 Using the histogram to measure area fraction: (a) photograph of a beef roast, after some image processing to enhance visibility of the fat and bones; (b) histogram of just the portion of the image containing the roast. The plot shows the number of pixels with each possible shade of brightness; the threshold setting shown (vertical line) separates the dark meat from the lighter fat and bone shows that 71% of the roast (by volume) is meat. The histogram method does not provide any information about the spatial distribution of the fat (marbling, etc.), which will be discussed in Chapter 4.

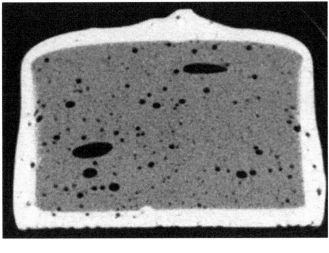

(a)

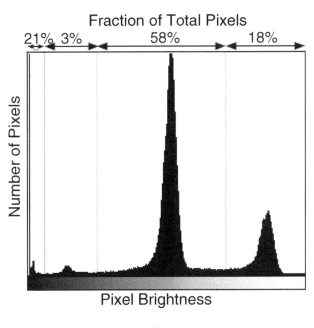

(b)

FIGURE 1.9 X-ray tomographic section through a Three Musketeers candy bar with its brightness histogram. The peaks in the histogram correspond to the holes, interior and coating seen in the image, and can be used to measure their volume fraction. (Courtesy of Greg Ziegler, Penn State University Food Science Department)

P_P (Equation 1.5), which is also a measure of the volume fraction. It is easy to see that as more and more points are placed in the 3D volume of the sample, that the point fraction must become the volume fraction.

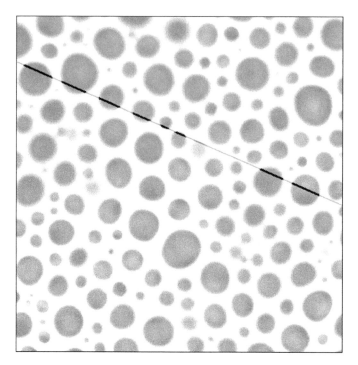

FIGURE 1.10 The image from Figure 1.2 with a random line superimposed. The sections that intersect pores are highlighted. The length of the highlighted sections divided by the length of the line estimates the volume fraction.

$$V_V = P_P \qquad (1.5)$$

The great advantage of a counting procedure over a measurement operation is not just that it is easier to make, but that the precision of the measurement can be predicted directly. If the sampled points are far enough apart that they act as independent probes into the volume (which in practice means that they are far enough apart then only rarely will two grid points fall onto the same portion of the structure being measured), then the counting process obeys Poisson statistics and the standard deviation in the result is simply the square root of the number of events counted.

In the grid procedure the events counted are the cases in which a grid point lies on structure being measured. So as an example, if a 49 point grid (7 × 7 array of points) is superimposed on the image in Figure 1.11, 16 of the points fall onto the bubbles. That estimates the volume fraction as 16/49 = 33%. The square root of 16 is 4, and 4/16 is 25%, so that is the estimate of the relative accuracy of the measurement (in other words, the volume fraction is reported as 0.33 ± 0.08). In order to achieve a measurement precision of 10%, it would be necessary to look at additional fields of view until 100 counts (square root = 10; 10/100 = 10%) had been accumulated. Based on observing 16 counts on this image, we would anticipate needing a total of 6 fields of view to reach that level of precision. For 5%, 400 counts are needed, and so forth.

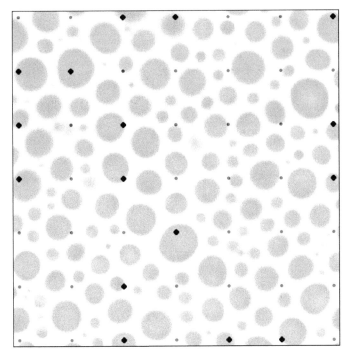

FIGURE 1.11 The image from Figure 1.2 with a 49 point (7 × 7) grid superimposed (points are enlarged for visibility). The points that lie on pores are highlighted. The fraction of the points that lie on pores estimates their volume fraction.

A somewhat greater number of points in the measurement grid would produce more hits. For example, using a 10 × 10 array of points on Figure 1.2 gives 33 hits, producing the same estimate of 33% for the volume fraction but with a 17% relative error rather than 25%. But the danger in increasing the number of grid points is that they may no longer be independent probes of the microstructure. A 10 × 10 grid comes quite close to the same dimension as the typical size and spacing of the bubbles. The preferred strategy is to use a rather sparse grid of points and to look at more fields of view. That assures the ability to use the simple prediction of counting statistics to estimate the precision, and it also forces looking at more microstructure so that a more representative sample of the whole object is obtained.

Another advantage of using a very sparse grid is that it facilitates manual counting. While it is possible to use a computer to acquire images, process and threshold them to delineate the structure of interest, generate a grid and combine it logically with the structure, and count the points that hit (as will be shown in Chapter 4), it is also common to determine volume fractions manually. With a simple grid having a small number of points (usually defined as the corners and intersections in a grid of lines, as shown in Figure 1.12), a human observer can count the number of hits at a glance, record the number and advance to another field of view.

At one time this was principally done by placing the grid on a reticle in the microscope eyepiece. With the increasing use of video cameras and monitors the

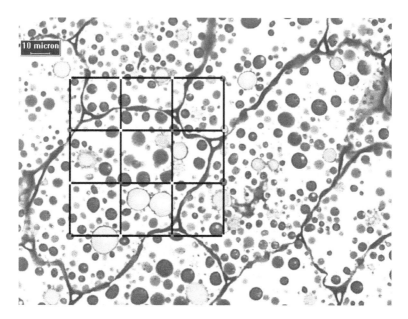

FIGURE 1.12 (See color insert following page 150.) A sixteen-point reticle randomly placed on an image of peanut cells stained with toluidine blue to show protein bodies (round, light blue) and starch granules (dark blue). The gaps at the junctions of the lines define the grid points and allow the underlying structure to be seen. Seven of the sixteen grid points lie on the starch granules (44%). The lines themselves are used to determine surface area per unit volume as described below (at the magnification shown, each line is 66 μm long). (Courtesy of David Pechak, Kraft Foods Technology Center)

same result can be achieved by placing the grid on the display monitor. Of course, with image capture the grid can be generated and superimposed by the computer. Alternately, printing grids on transparent acetate overlays and placing them on photographic prints is an equivalent operation.

By counting grid points on a few fields of view, a quick estimate of volume fraction can be obtained and, even if computer analysis of the images is performed subsequently to survey much more of the sample, at least a sufficiently good estimate of the final value is available to assist in the design of experiments, determination of the number of sections to cut and fields to image, and so on. This will be discussed a bit farther on. When a grid point appears to lie exactly on the edge of the structure, and it is not possible to confidently decide whether or not to count it, the convention is to count it as one-half.

This example of measuring volume fraction illustrates a trend present in many other stereological procedures. Rather than performing measurements of area or length, whenever possible the use of a grid and a counting operation is easier, and has a known precision that can be used to determine the amount of work that needs to be done to reach a desired overall result, for example to compare two or more types of material. Making measurements, either by hand or with a computer algorithm, introduces a finite source of measurement error that is often hard to estimate.

Even with the computer, some measurements, such as area and length, are typically more accurate than others (perimeter has historically been one of the more difficult things to measure well, as discussed in Chapter 5). Also, the precision depends on the nature of the sample and image. For example, measuring a few large areas or lengths produces much less total error than measuring a large number of small features. The counting approach eliminates this source of error, although of course it is still necessary to properly process and threshold the image so that the structure of interest is accurately delineated.

Volume fraction is an important property in most foods, since they are usually composed of multiple components. In addition to the total volume fraction estimated by uniform and unbiased (random) sampling, it is often important to study gradients in volume fraction, or to measure the individual volume of particular structures. These operations are performed in the same way, with just a few extra steps.

For example, sometimes it is practical to take samples that map the gradient to be studied. This could be specimens at the start, middle and end of a production run, or from the sides, top and bottom, and center of a product produced as a flat sheet, etc. Since each sample is small compared to the scale of the expected non-uniformities, each can be measured conventionally and the data plotted against position to reveal differences or gradients.

In other cases each image covers a dimension that encompasses the gradient. For instance, images of the cross section of a layer (Figure 1.13) may show a variation in the volume fraction of a phase from top to bottom. An example of such a simple vertical gradient could be the fat droplets settling by sedimentation in an oil and water emulsion such as full fat milk. Placing a grid of points on this image and recording the fraction of the number of points at each vertical position in the grid provides data to analyze the gradient, but since the precision depends on the number of hits, and this number is much smaller for each position, it is usually necessary to examine a fairly large number of representative fields to accumulate data adequate to show subtle trends.

Gradients can also sometimes be characterized by plotting the change of intensity or color along paths across images. This will be illustrated in Chapter 5. The most difficult aspect of most studies of gradients and nonuniformities is determining the geometry of the gradients so that an appropriate set of measurements can be made. For example, if the size of voids (cells) in a loaf of bread varies with distance from the outer crust, it is necessary to measure the size of each void and its position in terms of that distance. Fortunately, there are image processing tools (discussed in Chapter 4) that allow this type of measurement for arbitrarily shaped regions.

For a single object, the Cavalieri method allows measurement of total volume by a point count technique as shown in Figure 1.14. Ideally, a series of section images is acquired at regularly spaced intervals, and a grid of points placed on each one. Each point in the grid represents a volume, in the form of a prism whose area is defined by the spacing of the grid points and whose length is the spacing of the section planes. Counting the number of points that hit the structure and multiplying by the volume each one represents gives an estimate of the total volume.

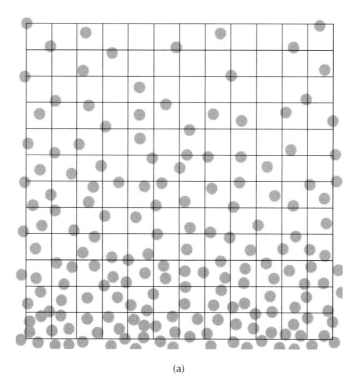

(a)

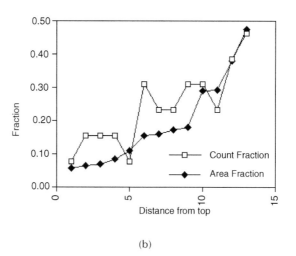

(b)

FIGURE 1.13 Diagram of a simple vertical gradient with a superimposed grid. Counting the fraction of the grid points (b) measures the variation with position, but plotting the area fraction provides a superior representation from one image.

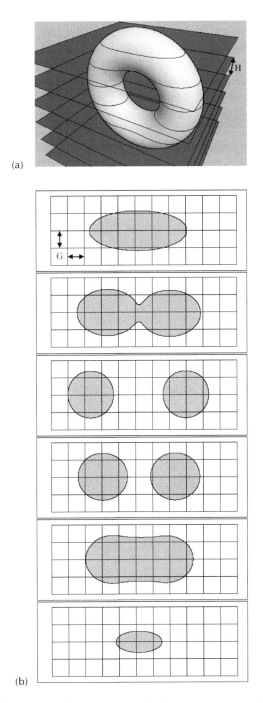

FIGURE 1.14 Illustration of the Cavalieri method for measuring an object's volume. A series of sections is cut with spacing $= H$ and examined with a grid of spacing G. The number of points in the grid that touch the object are counted (N). The volume is then estimated as $N \cdot H \cdot G \cdot G$.

SURFACE AREA

Besides volume, the most obvious and important property of three dimensional structures is the surfaces that are present. These may be surfaces that bound a particular phase (which for this purpose includes void space or pores) and separate it from the remainder of the structure which consists of different phases, or it may be a surface between two identical phase regions, consisting of a thin membrane such as the liquid surfaces between bubbles in the head on beer. Most of the mechanical and chemical properties of foods depend in various ways on the surfaces that are present, and it is, therefore important to be able to measure them.

Just as volumes in 3D structures are revealed in 2D section images as areas where the section plane has passed through the volume, so surfaces in 3D structures are revealed by their intersections with the 2D image plane. These intersections produce lines (Figure 1.15). Sometimes the lines are evident in images as being either lighter or darker than their surroundings, and sometimes they are instead marked by a change in brightness where two phase volumes meet. Either way, they can be detected in the image either visually or by computer-based image processing and used to measure the surface area that is present.

The length of the lines in the 2D images is proportional to the amount of surface area present in 3D, but there is a geometric factor introduced by the fact that the section plane does not in general intersect the surface at right angles. It has been

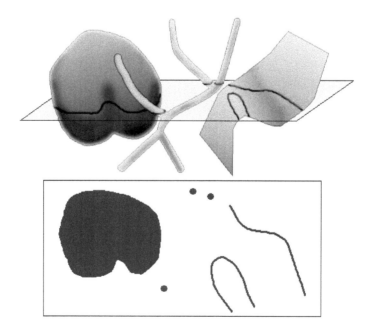

FIGURE 1.15 Passing a section plane through volumes, surfaces, and linear structures produces an image in the plane in which the volumes are shown are areas, the surfaces as lines, and the linear structures as points.

shown by stereologists that by averaging over all possible orientations, the mathematical relationship is

$$S_V = \frac{4}{\pi} \cdot B_A \qquad (1.6)$$

where S_V is the area of the surface per unit volume of sample and B_A is the length of boundary line per unit area of image, where the boundary line is the line produced by the intersection of the three-dimensional surface and the section plane. The geometric constant $(4/\pi)$ compensates for the variations in orientation, but makes the tacit assumption that either the surfaces are isotropic — arranged so that all orientations are equally represented — or that the section planes have been made isotropic to properly sample the structure if it has some anisotropy or preferred orientation.

This last point is critical and often insufficiently heeded. Most structures are not isotropic. Plants have growth directions, animals have oriented muscles and bones, manufactured and processed foods have oriented structures produced by extrusion or shear. Temperature or concentration gradients, or gravity can also produce anisotropic structures. This is the norm, although at fine scales emulsions, processed gels, etc. may be sufficiently isotropic that any orientation of measurement will produce the same result. Unless it is known and shown that a structure is isotropic it is safest to assume that it is not, and to carry out sampling in such a way that unbiased results are obtained. If this is not done, the measurement results may be completely useless and misleading.

Much of the modern work in stereology has been the development of sampling strategies that provide unbiased measurements on less than ideal, anisotropic or nonuniform structures. We will introduce some of those techniques shortly.

Measuring the length of the line in a 2D image that represents the intersection of the image plane with the surface in three dimensions is difficult to do accurately, and in any case we would prefer to have a counting procedure instead of a measurement. That goal can be reached by drawing a grid of lines on the image and counting the number of intersections between the lines that represent the surface and the grid lines. The number of intersection points per length of grid line (P_L) is related to the surface area per unit volume as

$$S_V = 2 \cdot P_L \qquad (1.7)$$

The geometric constant (2) compensates for the range of angles that the grid lines can make with the surface lines, as well as the orientation of the sample plane with the surface normal. This surprisingly simple-appearing relationship has been rediscovered (and republished) a number of times. Many grids, such as the one in Figure 1.12, serve double duty, with the grid points used for determining volume fraction using Equation 1.5, while the lines are used for surface area measurement using Equation 1.7.

As an example of the measurement of surface area, Figure 1.16 shows an image in which the two phases (grey and white regions, respectively) have three types of interfaces — that between one white cell and another (denoted α–α), between one grey cell and another (β–β), and between a white and a grey cell (α–β). The presence of many different phases and types of interfaces is common in food products.

By either manual procedures or by using the methods of image processing discussed in subsequent chapters, the individual phases and interfaces can be isolated and measured, grids generated, and intersections counted. Table 1.2 shows the

TABLE 1.2
Surface Area Measurements from Figure 1.16

Boundary Type	Intersection Counts	Cycloid length (μm)	$S_V = 2 \cdot P_L$ (μm^{-1})	Boundary Length (μm)	Image Area (μm^2)	$S_V = 4/\pi \cdot B_A$ (μm^{-1})
α–α	24	360	0.133	434.8	4500	0.123
α–β	29	360	0.161	572.5	4500	0.162
β–β	9	360	0.050	117.8	4500	0.033

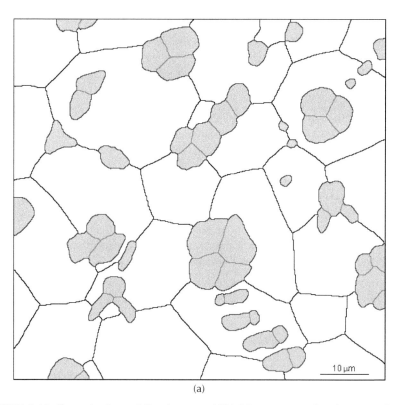

(a)

FIGURE 1.16 (See color insert following page 150.) Measurement of surface area. A two-phase microstructure is measured by (a) isolating the different types of interface (shown in different colors) and measuring the length of the curved lines; and (b) generating a cycloid grid and counting the number of intersections with each type of interface. The reason for using this particular grid is discussed in the text.

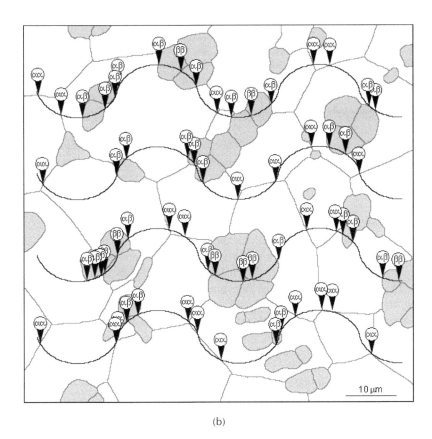

(b)

FIGURE 1.16 (continued)

specific results from the measurement of the length of the various boundary lines
and from the use of the particular grid shown. The numerical values of the results
are not identical, but within the expected variation based on the precision of the
measurements and sampling procedure used.

Note that the units of P_L, B_A and S_V are all the same (length^{-1}). This is usually
reported as (area/volume), and to get a sense of how much surface area can be
packed into a small volume, a value of 0.1 μm^{-1} corresponds to 100 cm^2/cm^3, and
values for S_V substantially larger than that may be encountered. Real structures often
contain enormous amounts of internal surface within relatively small volumes.

For measurement of volume fraction the image magnification was not important,
because P_P, L_L, A_A and V_V are all dimensionless ratios. But for surface area it is
necessary to accurately calibrate image magnification. The need for isotropic sampling is still present, of course. If the section planes have been cut with orientations
that are randomized in three dimensions (which turns out to be quite complicated
to do, in practice), then circles can be drawn on the images to produce isotropic
sampling in three dimensions.

One approach to obtaining isotropic sampling is to cut the specimen up into many small pieces, rotate each one randomly, pick it up and cut slices in some random orientation, draw random lines on the section image, and perform the counting operations. That works, meaning that it produces results that are unbiased even if the sample is not isotropic, but it is not very efficient. A better method, developed nearly two decades ago, generates an isotropic grid in 3D by canceling out one orientational bias (produced by cutting sections) with another. It is called the method of "vertical sections" and requires being able to identify some direction in the sample (called "vertical" but only because the images are usually oriented with that direction vertical on the desk or screen). Depending on the sample, this could be the direction of extrusion, or growth, or the backbone of an animal or stem of a plant. The only criterion is that the direction be unambiguously identifiable. The method of vertical sections was one of the first of the developments in what has become known as "unbaised" or design-based stereology.

Section planes through the structure are then cut that are all parallel to the vertical direction, but rotated about it to represent all orientations with equal probability (Figure 1.17). These planes are obviously not isotropic in three-dimensional space, since they all include the vertical direction. But lines can be drawn on the section plane images that cancel this bias and which are isotropic. These lines must have sine-weighting, in other words they must be uniformly distributed over directions based not on angles but on the sines of the angles, as shown in the figure. It is possible to draw sets of straight lines that vary in this way, but the most efficient procedure to draw lines that are also uniformly distributed over the surface is to generate a set of cycloidal arcs.

The cycloid is a mathematical curve generated by rolling a circle along a line and tracing the path of a point on the rim (it can be seen as the path of a reflector on a bicycle wheel, as shown in Figure 1.18). The cycloid is exactly sine weighted and provides exactly the right directional bias in the image plane to cancel the orientational bias in cutting the vertical sections in the first place. Cycloidal arcs can be generated by a computer and superimposed on an image. The usual criteria for independent sampling apply, so the arcs should be spaced apart to intersect different bits of surface line, and not so tightly curved that they resample the same segment multiple times. They may be drawn either as a continuous line or separate arcs, as may be convenient. Figure 1.18 shows some examples.

The length of one cyloidal arc (one fourth of the full repeating pattern) is exactly twice its height (which is the diameter of the generating circle), so the total length of the grid lines is known. Counting the intersections and calculating S_V using Equation 1.7 gives the desired measure of surface area per unit volume, regardless of whether the structure is actually isotropic or not. Clearly the cutting of vertical sections and drawing of cycloids is more work than cutting sections that are all perpendicular to one direction (the way a typical microtome works) and using a simple straight-line grid to count intersections. Either method would be acceptable for an isotropic structure, but the vertical section method produces unbiased results even if the structure has preferred orientation, and regardless of what the nature of that anisotropy may be.

(a)

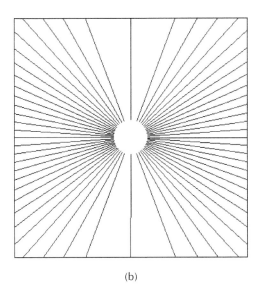

(b)

FIGURE 1.17 The method of vertical sections: (a) a series of slices are cut lying parallel to an identifiable direction, but rotated to different angles about that direction; (b) on each slice, lines that are sine-weighted (their directions incremented by equal steps in the value of the sine of the angle) are drawn. These lines isotropically sample directions in three dimensional space.

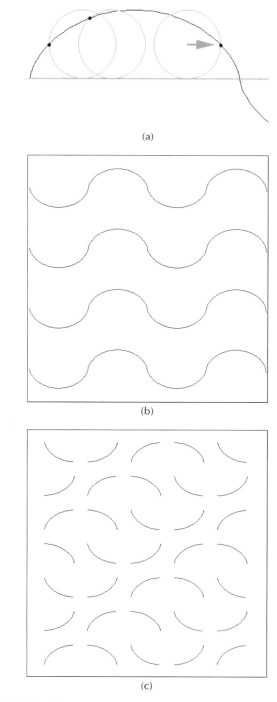

(a)

(b)

(c)

FIGURE 1.18 Cycloid grids: (a) generation of a cycloid as the path traced out by a point on the rim of a rolling circle; (b) a set of continuous cycloid grids; (c) a set of separated cycloid arcs.

LINES AND POINTS

The preceding sections have described the measurement of volumes and surfaces that may be present in 3D structures. These are, respectively, 3- and 2-dimensional features. There may also be 1- and 0-dimensional features present, namely lines and points. Surfaces were considered to include extremely thin interfaces between phases, as well as finite membranes around objects. Similarly, a linear structure may have finite thickness as long as its lateral dimensions are very small compared to its length and the size of the other structures with which it interacts. So the veins or nerves in meat, and the various kinds of fibers in either natural or man-made foods are all linear structures.

A thicker structure, such as the network of particles that form in gels (e.g., polysaccharides such as pectin or alginates), shortening and processed meats, may also be considered as a linear structure for some purposes, as can a pore network. In both cases, we imagine the lateral dimensions to shrink to form a backbone or skeleton of the network, which is then treated as linear for purposes of measurement. Note that a linear structure may consist of a single long line, many short ones, or a complex branching network. The topology of structures is considered later, at the moment only the total length is of concern.

In addition, a line exists where two surfaces meet, as indicated in Figure 1.19. One of the simplest examples of these edge lines is the structure of a bubble raft such as the head on beer. Except for the bubbles on the outside of this raft, whose surfaces are curved, all of the soap films that separate bubbles from each other are flat planes. This is the equilibrium structure of many solid materials as well, ranging from grains in metals to cells in plants. The boundaries of each facet where three planes meet are lines, and can be treated as an important component of the structure. It is these triple lines where much of the diffusion of gases and fluids occurs, for example, or which are responsible for the mechanical strength of a fiber network.

Linear structures appear as points in a section plane, where the plane intersects the line. In many real cases the lateral dimension of the linear structure is small but still large enough that the intersections appear as small features in the image. These

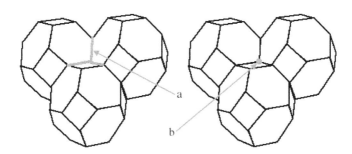

FIGURE 1.19 Diagram of a cell or bubble structure (with the topmost cell removed for clarity) showing the triple lines (a) where three cells meet and the quadruple points (b) where four cells meet.

are simply counted. The total length of the linear structure per unit volume L_V of material is calculated from the number of intersection points per unit area P_A as

$$L_V = 2 \cdot P_A \qquad (1.8)$$

where P is the number of points of intersection, A is the area of the image, and (2) is a geometrical constant, as above, that compensates for the range of orientations with which the section plane can intersect the lines.

As for the measurement of surfaces, discussed above, the measurement of line length must be concerned with directionality. If the sample is not isotropic, then the probes must be. In this case the probes are the section planes, and it was noted above that producing an isotropic array of section planes is very difficult, inefficient, and wasteful of material. There is a useful technique that can use the method of vertical sections to simplify the procedure.

Thus far, measurements have been made on plane sections cut through surfaces. Either the material has been considered as opaque so that a true plane surface is examined, or in the case of transmission microscopy, the section thickness has been assumed to be very thin as compared to the dimensions of any of the structures of interest. But in many cases the food products of interest are at least somewhat transparent and it is possible to obtain images by shining radiation (light, electrons, or something more exotic) through a moderately thick slice. The resulting image shows a projection through the structure in which linear features can be seen.

Simply measuring the length of the lines will not suffice, however. There is no reason to expect them to all lie flat in the plane of the section, so that their true length can be measured, and there is likewise no reason to expect them to be isotropic in direction so that a geometric constant can be used to convert the total measured projected length to an estimate of the true length in 3D.

But another approach is possible. Imagine drawing a line on the image. That line represents a plane in the original thick slice sample that extends down through the thickness of the slice, as shown in Figure 1.20. Counting the number of inter-sections of the linear structure with the drawn line (which implies their intersection with the plane the line represents) gives a value of P_A (number of counts per unit area of plane). The area of the plane is just the length of the line drawn on the image times the thickness of the section, which must be known independently. Then the same relationship introduced above (Equation 1.8) can be used.

In order to obtain isotropic orientation of the plane probes (not the section planes, but the thru-the-slice planes that correspond to the lines drawn on the image), it is necessary to use the vertical sectioning approach. All of the slices are cut parallel to some assumed vertical orientation and rotated about it. Then the lines are drawn as cycloids, representing a cycloidal cylindrical surface extending down through the section thickness. Because in this case it is the orientation of the surface normals of those probe surfaces that must be made isotropic, it is necessary to rotate the grid of cycloid lines by 90 degrees on the image, as shown in Figure 1.21.

Finally, there are many structures in which the features of interest are small in dimension as compared to the scale of the image and of surrounding structures.

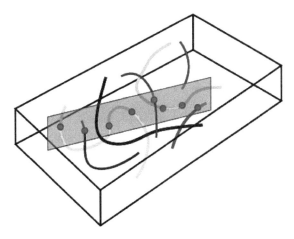

FIGURE 1.20 Linear structures in a thick section can be measured by counting the number of intersections they make with a plane extending through the section thickness, represented by a line drawn on the projected image.

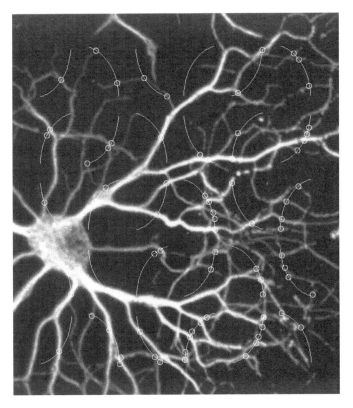

FIGURE 1.21 Measuring the length per unit volume of tubules in a thick section, by drawing a cycloid grid and counting the number of intersections.

These are typically called "points." Points also mark the locations where lines meet planes, such as a linear structure penetrating a boundary, or even four planes meeting (the quadruple points that exist in a tessellation of grains or cells where four cells meet). A section plane through the structure will not intersect any true points, so it is not possible to learn anything about them from a plane or thin slice. But if they are visible in the projected transmission image from a thick slice they can be counted. Provided that the number of points is low enough (and they are small enough) that their images are well dispersed in the image, then simple counting with produce P_V directly, where V is the volume (area times thickness) of the slice imaged, and P is the observed number of points.

DESIGN OF EXPERIMENTS

Given a specimen, or more typically a population of them, what procedure should be followed to assure that the measurement results for area fraction, surface area per unit volume, or length per unit volume are truly representative? In most situations this will involve choosing which specimens to cut up, which pieces to section, which sections to examine, where and how many images to acquire, what type of grid to draw, and so forth. The goal, simply stated but not so simply achieved, is to probe the structure uniformly (all portions equally represented), isotropically (all directions equally represented) and randomly (everything has an equal probability of being measured). For volume or point measurements, the requirement for isotropy can be bypassed since volumes and points, unlike surfaces and lines, have no orientation, but the other requirements remain.

One way to do this, alluded to above, is to achieve randomization by cutting everything up into little bits, mixing and tumbling them to remove any history or location or orientation, and then select some at random, microtome or section them, and assume that the sample has been thoroughly randomized. If that is the case, then any kind of grid can be used that doesn't sample the microstructure too densely, so that the locations sampled are independent and the relationship for precision based on the number of events counted holds. To carry the random approach to its logical conclusion, it is possible to draw random lines or sprinkle random points across the image. In fact, if the structure has some regularity or periodicity on the same scale as the image, a random grid is a wise choice in order to avoid any bias due to encountering a beat frequency between the grid and the structure.

But random methods are not very efficient. First, the number of little bits and the number of sections involved at each step needs to be pretty large to make the lottery drawing of the ones to be selected sufficiently random. Second, any random scheme for the placement of sections, fields of view or grids, or the selection of samples from a population, inevitably produces some clustering that risks oversampling of regions combined with gaps that undersample other areas.

The more efficient method, systematic (or structured) random sampling, requires about one-third as much work to achieve comparable precision, while assuring that IUR (Isotropic, Uniform, Random) sampling is achieved. It starts, as does any design of experiments, by making a few preliminary measurements to allow estimating the number of samples, sections, images, etc. that will be required to achieve the desired precision.

Let us say by way of example that we wish to measure the amount of surface area per unit volume in a product, for which we have 12 specimens. After considering the need to compare these results to others of products treated differently, we decide that the results needs to be measured to 3% precision. That means we need at least 1000 events (hits made by the sampling line probes — the grid — with the lines in the image that represent the surface in the 3D structure) for each population, because the square root of 1000 is about 32, or 3%.

To get those 1000 hits, we could cut a single section from a single specimen and measure many fields of view with a grid containing many lines, but it seems very dangerous to assume that one individual specimen is perfectly representative of the population, or that one section orientation is an adequate sampling of the structure. But on the other hand, we certainly want to look at as few specimens as possible and cut as few sections as practical.

Examining a few randomly selected images (or perhaps prior experience with other similar products) suggests that with a fairly sparse grid (so that there is no danger of oversampling), a typical field of view will produce about 10 hits. That means we need to look at no fewer than 100 fields of view. How should these be distributed over the available samples?

The variables we can choose are the number of specimens to cut up (N), the number of vertical section orientations to cut in each (V), the number of slices to cut at each orientation (S), and the number of fields of view to image on each slice (F). The product of $N \cdot V \cdot S \cdot F$ must be at least 100. There is a different cost to each variable, with F being much quicker, cheaper and easier than V, for example. And some variables have natural limits — clearly N must be larger than 1 but no greater than 12, and it is usual to select V as an odd number, say 3 or 5, rather than an even one in order to avoid any effects of symmetry in the sample. This is perhaps slightly more important for natural products, which tend to grow with bilateral symmetry, than man-made ones.

One choice that could be made would be $N = 3$, $V = 3$, $S = 4$, $F = 3$ (a total of 108 fields). Since we observe (in this example) that about half of each slide actually contains sample, and half is empty, we will double the number of fields to $F = 6$ to compensate. Other choices are also possible, but usually the various factors will tend to be of similar magnitudes. Based on this choice, how should the three specimens, location of the various fields of view, etc., be carried out?

That is where the systematic random part of the procedure comes in. We want uniform but random coverage, meaning that every possible field of view in every possible slice in every possible orientation of every possible specimen has an equal chance of being selected, even if only a few of them actually will be. The procedure is the same at every stage of the selection process. Starting with the 12 specimens in the population, we must choose $N = 3$. Twelve divided by three is four, so we begin by generating a random number (in the computer or by spinning a dial) between 1 and 4. The specimen with that number is selected, and then every fourth one after it. The basic principle is shown in Figure 1.22. This procedure distributes the selection uniformly across the population but randomizes the placement.

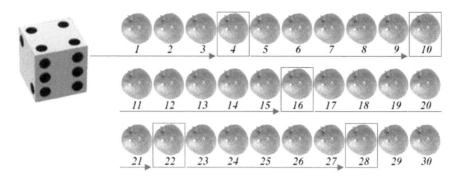

FIGURE 1.22 The principle of systematic random selection. Consider the task of selecting 5 apples from a population of 30. Dividing 5 into 30 gives 6, so generate a random number from 1 to 6 to select the first specimen. Then step through the population taking every sixth individual. The result is random (every specimen has an equal probability of being selected) and uniform (there are no bunches or gaps in the selection).

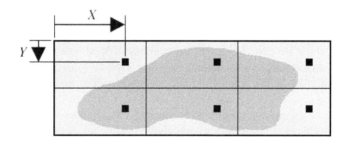

FIGURE 1.23 Systematic random location of fields of view on a slide. The total area of the slide is subdivided into identical regions, six in this example, corresponding to the number of fields of view to be imaged. Two random numbers are generated for the X and Y coordinates of the first field of view in the first region. The remaining fields of view come from the identical positions within the other regions. Different random placement is used for each slide.

Similarly, the rotation angles around the vertical direction can be systematically randomized. If three orientations are to be used, 360/3 = 120 degrees. So a random number from 1 to 120 is generated and used for the initial angle of rotation, and then 120 degree steps are used to orient the blocks that will be sectioned for the other two directions. For the four sections to be examined from each block, it is necessary to know how many total sections could be cut from each. For the purposes of illustration, assume that this is 60. Sixty divided by four equals 15, so a random number from 1 to 15 selects the first section to be examined, and then the fifteenth, thirtieth and forty-fifth ones past that are chosen.

For the $F = 6$ fields of view, the total area of the cut section or slide is divided into 6 rectangles. Two random numbers are needed to specify coordinates of a field of view in the first rectangle. Then the same locations are used in the remaining rectangles for the additional fields of view. Figure 1.23 illustrates the method of systematic random location of fields of view on a slide. This method can be generalized to handle any other sampling requirement, as well, such as the placement of grid lines or points, and so forth.

TOPOLOGICAL PROPERTIES

The volume of 3D structures, area of 2D surfaces, and length of 1D lines, can all be measured from section images. The most efficient methods use counting procedures with a point grid for volumes or a line grid for surfaces. The results of these procedures, with a little trivial arithmetic, give the metric properties of the structure. Of course, there may be many components of a typical food structure, including multiple phases whose volumes can be determined, many different types of surfaces (this includes potentially interfaces between each of the phases in the material, but usually most of these combinations are not actually present), and all sorts of linear structures. But with appropriate image processing to isolate each class of structure, they can all be measured with the procedures described.

Consider the case in Figure 1.24. From the various sections through the structure the total volume, the surface area, and the length of the tubular structure can be determined. But the fact that the tube is a single object, not many separate pieces, that it is tied into a knot, and that the knot is a right-handed overhand knot, is not evident from the individual section images. It is only by combining the section images, knowing their order and spacing, and reconstructing the 3D appearance of the structure that these topological properties appear.

Topological properties of structures are often as important as the metric ones, but require a volume probe rather than a plane, line or point probe as used for the metric properties. The simplest and most familiar topological property is simply a number, such as the number of points or features counted in a volume as described above. The thick slice viewed in transmission is one kind of volume probe. Another is a full three-dimensional imaging method, such as serial section reconstruction, or magnetic resonance or CT imaging. In some cases, such as knowing that the knot is right handed, full 3D imaging is necessary, but fortunately there is another easier procedure that can usually provide basic 3D topological information.

The simplest volume probe consists of two parallel section images, a known distance apart. These are compared to detect events that lie between them, which means that they must be close enough together that nothing can happen that is not interpretable from the comparison. In general, that means the plane separation should be no more than one fourth to one fifth of the size of the features that are of interest in the structure.

The simplest kind of disector measurement (Figure 1.25) works to determine the number per unit volume of convex, but arbitrarily shaped features. Each such feature must have one, and only one, lowest point. Counting these points counts the features, and they can be detected by observing all features that intersect one of the two sections and thus appear in one image, but do not intersect the second section and, hence, are not seen in that image. N_V (number per unit volume) is measured by the number of these occurrences divided by the volume examined, which is the product of the image area times the distance between the sections.

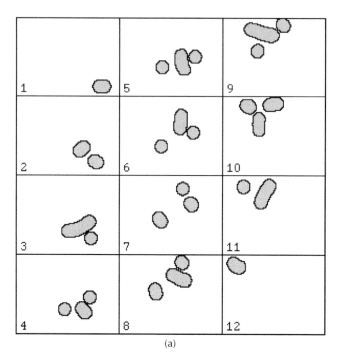

(a)

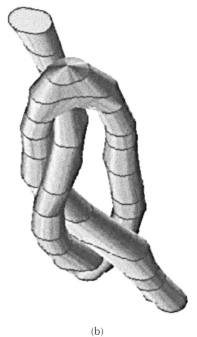

(b)

FIGURE 1.24 Twelve slices through a structure from which the metric properties can be determined, and a three-dimensional reconstruction from the sections which reveals the topological properties.

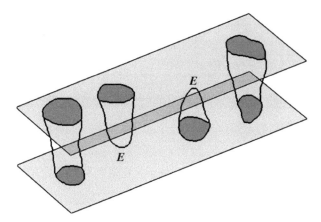

FIGURE 1.25 Diagram of disector logic. Features that appear in either section plane that are absent from the other represent ends (marked E). Features that are matched are not counted.

The efficiency and precision of this procedure can be improved by counting tops as well as bottoms, and dividing by two. So the features that appear in either section image but are NOT matched in the other image are counted, and

$$N_V = \frac{E}{2 \cdot Area \cdot Distance} \qquad (1.9)$$

where E is the number of ends.

We will see in Chapter 5 (Figure 5.7) that there are image processing techniques that make it fairly quick to count the features that are not matched between the two images. This is important because if the spacing between the planes is small (as it must be), most of the features that intersect one plane will also intersect the second, and it becomes necessary to examine a fairly large area of images (which must be matched up and aligned) in order to obtain a statistically useful number of counts.

Because it is a volume probe and counts points, neither of which have any directionality associated with them, the disector does not require isotropy in its placement. Only the simpler requirements of uniformity and randomness are needed, and so sectioning can be performed in any convenient orientation.

With the disector it is possible to overcome the limitations discussed previously of assuming a known size or shape for particles in order to determine their number or mean size. For example, the pores or cells in bread are clearly not spherical and certainly not the same size or shape. The disector can count them to determine number density N_V. Figure 1.26 shows one simple way to do this: to cut a very thin slice (e.g., about 1 mm thick) and count the number of pores that do NOT extend all the way through. This can be facilitated by a little sample preparation. Applying a colored ink to the top surface of the slice with an ink roller makes it easy to see the pores, and placing the slice on a different color surface makes it easy to see those that extend clear through.

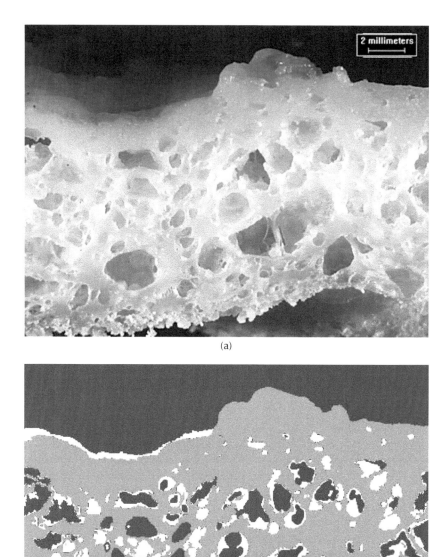

(a)

(b)

FIGURE 1.26 (See color insert following page 150.) Application of the disector to counting the number of pores in a baked product: (a) photograph of one cut surface; (b) result of inking the cut surface and placing a thin slice on a contrasting color background. (Courtesy of David Pechak, Kraft Foods Technology Center)

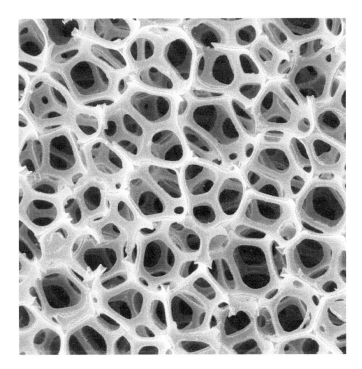

FIGURE 1.27 Example of a network that formed along the triple lines between bubbles in a foam, showing a high connectivity.

Since this disector counts only bottoms of pores, and not their tops, the number per unit volume is just the number of pores divided by the volume of the slice (area times thickness). But there is more information available. The volume fraction of the pores is readily measured by the area fraction of the top surface, for instance by counting the number of inked pixels and dividing by the total area of the slice, or if desired by applying a point grid and dividing the number that fall on the inked regions to those that fall anywhere on the slice. Knowing the volume fraction of pores and the number per unit volume allows a simple calculation of the mean volume of a pore, without any assumptions about shape.

The disector method can be extended quite straightforwardly to deal with features that may not be convex, and even to characterize the topological properties of networks. The key property of a network is its connectivity, or the genus of the network. This is the number of redundant connections that exist, paths that could be cut without separating the network into two or more pieces. Figure 1.27 shows one of the many types of high-connectivity networks that can exist.

The network in Figure 1.27 formed along the triple lines where bubbles met, producing a very open and regular structure that can be visually comprehended from SEM images, and can be characterized in several ways including the size of the original bubbles and the lengths of the connections in the network. When the network structure is more irregular and dense, it is more difficult to visualize it using the SEM. Figure 1.28 shows an example. Even serial section reconstruction and visualization of this

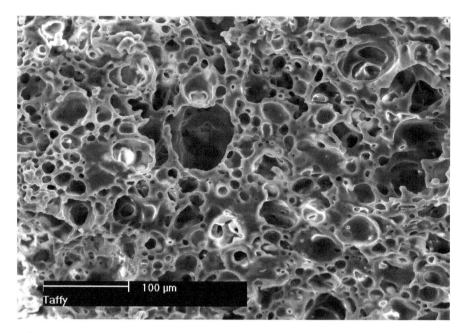

FIGURE 1.28 SEM image of the network structure in highly aerated taffy. (Courtesy of Greg Ziegler, Penn State University Food Science Department)

type of structure does not convey a view that is very instructive and certainly is not suitable for quantitative comparisons. The disector approach using two parallel section images a known distance apart to count topological events is perhaps the only way to gain a measurement that can be meaningfully used to correlate structural changes with processing or performance variables. Of course, the section images can also be used to determine the volume fraction of the solid material.

Counting the ends that occur between the two section images provides a way to count the number of convex features per unit volume, regardless of shape. To count features that are branched and complex in shape, or to learn about the connectivity of networks, two additional types of events must also be identified and counted. The end points as described above are now referred to as T^{++} points, meaning that they are convex tangent counts. If an end occurs between the planes, then there must be a point where a tangent plane parallel to the section planes just touches the end of the feature, and at this point the curvature of the feature must be convex (the radii of curvature are inside the feature).

As shown in Figure 1.29, there are two other possibilities. A negative or T^{--} tangent count occurs when a hole within a feature ends between the planes. In appearance this corresponds to a hollow circle in one plane with a matching filled-in circle in the other. The tangent plane at the end of this hollow will locate a point where both radii of curvature lie outside the body of the feature (and in the hollow). These events are typically rather rare.

The other type of event is a branching. If a single intersection in one plane splits to become two in the second plane it implies that between the two section planes

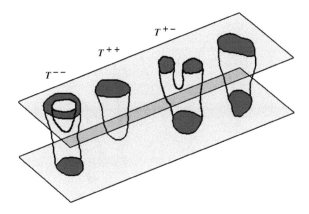

FIGURE 1.29 Diagram of extended disector logic. Features with ends between the section planes are convex and correspond to positive tangent counts (T^{++}). Pores that end are concave and are labeled as negative tangent counts (T^{--}), while branches have saddle curvature and are labeled as mixed tangent counts (T^{+-}).

there must be a point where the tangent plane would find saddle curvature — a point where the two principal radii of curvature lie on opposite sides of the surface. Hence, this is called a "mixed tangent" or "saddle point" and denoted by T^{+-}.

Counting these events allows the calculation of the net tangent count per unit volume or T_V, where as before the volume is the product of the area examined and the separation distance between the planes. The net tangent count is just (T^{++}) + (T^{--}) – (T^{+-}), and the Euler characteristic is one half of the net tangent count. But the Euler characteristic is also $N_V - C_V$, the difference between the number of discrete (but arbitrarily shaped) features per unit volume and the connectivity per unit volume.

$$N_V - C_V = \frac{T^{++} + T^{--} - T^{+-}}{2 \cdot Area \cdot Distance} \tag{1.10}$$

Many real samples of interest consist of discrete features that may be quite irregular in shape with complex branching and protrusions. For a set of dispersed features that are irregular in shape, the connectivity C_V is zero. Consequently the number of features per unit volume is calculated from the net tangent count. Subtracting the mixed or saddle counts corrects for the branching in the feature shapes as indicated schematically in Figure 1.30.

Another type of structure that is often encountered is an extended network that may have many dead ends but is still continuous. In this case, $N_V = 1$. In the previous example of discrete features, the multiple branches and ends for a single feature may add to the T^{++} count, but the T^{+-} count must increase along with it, as shown in Figure 1.30. The net tangent count will still be two and the number of features one. But in the case of a network, the T^{+-} count will dominate, even though local ends of network branches may produce some T^{++} events. In Equation 1.10, T^{+-} and

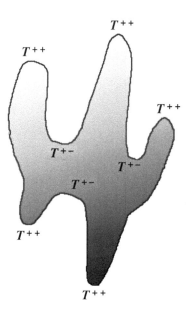

FIGURE 1.30 Tangent counts on a simply connected nonconvex feature; the net tangent count is still +2 (= +5 − 3 in this example) regardless of the number of branches and ends.

C_V both appear with minus signs, so the net abundance of tangent counts measures the network connectivity.

OTHER STEREOLOGICAL TECHNIQUES

This introductory chapter has covered the most commonly used stereological calculations, but the subject is quite large, and a variety of measurement methodologies have been derived for specific circumstances. There is also in the modern literature a wealth of information on sample preparation techniques that avoid bias, and specific approaches that are appropriate for newer microscopies including the confocal light microscope.

As a simple example of the latter, consider again the measurement of the total length of fibers per unit volume. The length can be measured by counting intersections with a surface, but as fibers have an orientation, we would like to use as a probe a surface that is anisotropic. One way to do this is to collect a set of parallel optical section images with the microscope and then place a set of circles on them that vary in diameter to correspond to a sphere in 3D space, as shown in Figure 1.31. Counting the number of intersections of the fibers with these lines allows an unbiased calculation of L_V, the length of the fibers, as $L_V = 2P_A$ where A is the surface area of the sphere.

There are also stereological models that apply to particular kinds of microstructures. One that is encountered with some frequency is a structure containing a layer, or a thick membrane, or any other structure that can be described as a "muralia," "plate," "sheet" or "blanket" in three-dimensional space. Measuring the thickness

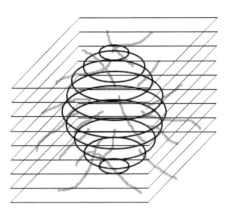

FIGURE 1.31 An isotropic sphere probe consists of circles in serial confocal sections that form a sphere. Counting the number of linear features that cross the various circles gives the number of counts P that can be combined with the surface area of the sphere A to calculate length per unit volume.

of that layer is important, but it may not be directly accessible in the section images cut through it at arbitrary angles. In general, as shown in the diagram, the apparent width of the section will be greater than the true three-dimensional thickness.

This is a typical case in which stereology tells us what should be measured. There are computer-based image analysis techniques that can measure the two-dimensional width of the section, but this is a case where they should not be used, as the result is meaningless. Instead, it is useful to consider the distribution of the intersection lengths of line probes that pass through a thick layer (Figure 1.32). The shape of this probability distribution, which falls off asymptotically as the intersection length increases, is hard to treat analytically. But replotting the data as the frequency distribution of the reciprocal of the intersection length offers a dramatic simplification — a straight line distribution. In real structures, the finite extent and curvature of the plates limits the number of extremely long intersections, which correspond to the very small vales of the reciprocal, but this has little effect on the distribution. The nice thing about this simple triangular shape is that the mean value is just two thirds of the maximum, and the maximum is the reciprocal of the true 3D thickness of the layer.

So a measurement procedure would be to draw lines in many orientations on the image and measure intercept lengths, to determine the average value of the reciprocal. Chapter 4 (Figure 4.21) shows an example in which computer-generated lines and automatic measurements are used. Note that these lines may be random lines, or we can use the systematic random method to rotate a grid of parallel lines to different orientations. Either way, the mean value of the inverse of the intercept length λ is determined. Then the true thickness of the layer t, which may not even be seen in any of the images, is calculated as

$$\frac{1}{t} = \frac{3}{2} \cdot \left(\frac{1}{\lambda} \right)_{mean} \tag{1.11}$$

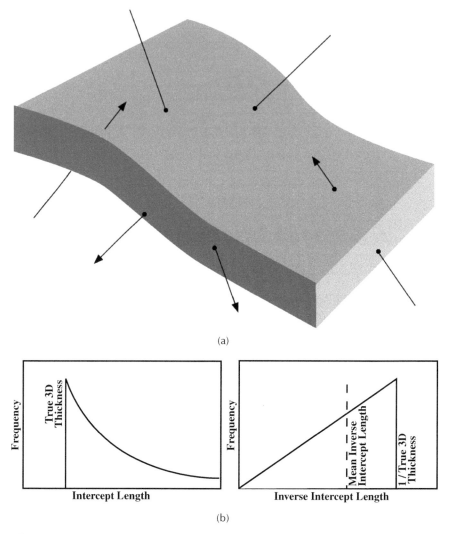

FIGURE 1.32 Diagram of random intercepts through a thick layer, showing frequency distributions of intercept lengths and the reciprocal of intercept lengths, and the relationship of those values to the true 3D dimensional thickness of the layer.

This procedure involves measuring the lengths of intercepts with a grid of parallel lines that is rotated, and is very much like another that can be used to characterize the anisotropy of a structure. We have seen ways to use vertical sectioning and cycloidal grids to avoid any bias in determining accurate values for S_V and L_V. However, since many food structures, both natural and man-made, exhibit preferred orientation to a greater or lesser degree, it is also interesting to find ways to characterize this anisotropy. One method is to represent it by carefully preparing section planes normal to each of the principal axes of the structure, which can usually be predicted based on consideration of the history or geometry of the subject. As shown

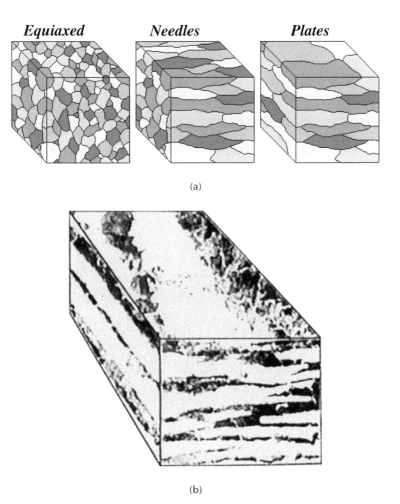

(a)

(b)

FIGURE 1.33 The appearance of anisotropic structures on perpendicular sections: (a) diagram of equiaxed, needle-like and plate-like cells; (b) three views of a natural layered product.

in Figure 1.33, it is not possible to understand or measure the structure without using more than one view. Many structures in food have entirely different appearances in different orientations. Such pictures as those in Figure 1.34 are only anecdotal and it is often useful to have numerical representations for comparison.

Creating a grid of parallel lines and applying it to the structure at different orientations uses the same procedure as the measurement of layer thickness, except that it is the mean value of the intercept length rather than its inverse that is wanted. For a space-filling arrangement of cells, the easiest way to get the mean value is to count intersections of the grid lines with the cell boundaries (mean intercept length = total line length/number of intersections). Plotting the mean intercept length vs. angle generates a "rose plot" or radial diagram that identifies the principal axis of the anisotropy and also its magnitude as shown in Figure 1.35.

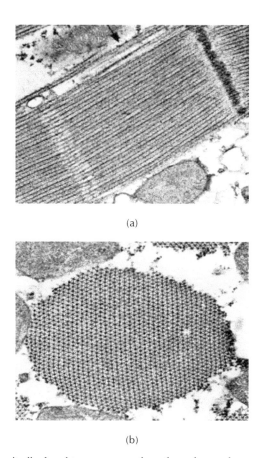

(a)

(b)

FIGURE 1.34 Longitudinal and transverse sections through muscle.

Intercept lengths are used in another way in a technique "known as point sampled intercepts." This can be used in conjunction with the disector method to determine more about the sizes of particles in 3D without making any shape assumptions. The method requires two steps. First, a grid of points is superimposed on the image. If the sample structure is very regular or repetitious, this grid should be random, but in most cases the local structure is sufficiently irregular to allow the use of a regular grid. The fraction of points that fall onto particles gives the volume fraction of the particles, which combined with a determination of number per unit volume, allows calculation of the mean particle size.

The second step is to use each point that lands on a particle to initiate a line. The line orientations should be isotropic and randomized, using the same procedures already discussed. If the section is a vertical section, the orientations should be sine-weighted. Each line extends from the initial point to the edge of the particle in the chosen direction as shown in Figure 1.36. The length of the intercept line is measured (r), and used to calculate a volume-weighted mean volume as the mean value of ($4/3 \cdot r^3$). The phrase volume-weighted means that the particles are sampled with a

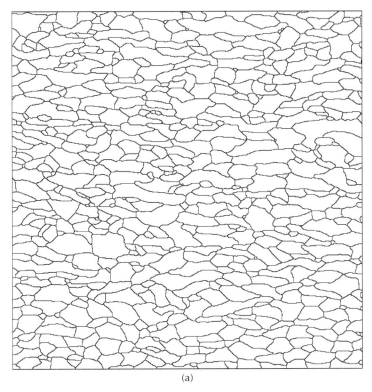

(a)

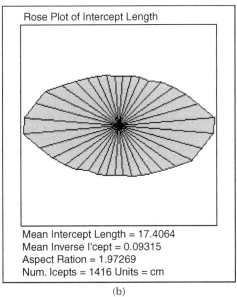

Rose Plot of Intercept Length

Mean Intercept Length = 17.4064
Mean Inverse I'cept = 0.09315
Aspect Ration = 1.97269
Num. Icepts = 1416 Units = cm

(b)

FIGURE 1.35 Measurement of anisotropy: (a) longitudinal section through cells in a plant stalk; (b) measurement results obtained by plotting the mean intercept length with a grid of parallel lines rotated in 10-degree steps.

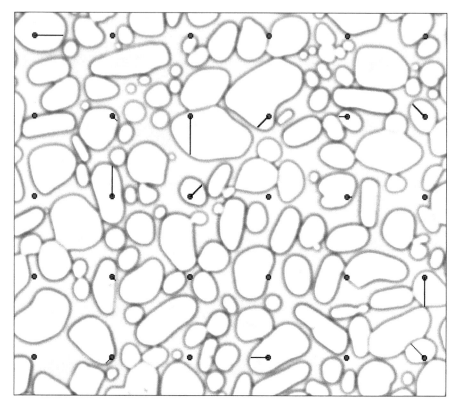

FIGURE 1.36 The method of point sampled intercepts applied to a confocal microscope image of partially melted ice crystals in ice cream. A grid of points is superimposed on the image, and from each point that lies within a crystal a line is drawn to measure the radial distance to the boundary as described in the text.

probability proportional to their volume (which controls the probability that a grid point will fall on the particle). The mean volume is an estimate of the particle volume from the radial distance from the random point to the particle boundary. Although the equation looks like that for the volume of a sphere, actually it is also a correct way to estimate particle volume for arbitrary shaped objects using random point-sampled intercepts.

A very useful stereological relationship uses this volume-weighted mean volume to determine the breadth of the distribution of the conventional ("number-weighted") volume. The variance (square of the standard deviation) of that distribution is the difference between the square of the mean volume (for instance calculated using the volume fraction and number per unit volume) and the volume-weighted mean volume. This is independent of particle shape.

Compare this methodology to the attempt to determine the distribution of particle sizes by assuming a spherical shape, measuring a distribution of circle sizes, and performing the unfolding operation described at the beginning of this chapter. That technique is mathematically suspect and critically dependent on a shape assumption,

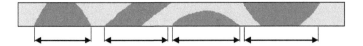

FIGURE 1.37 Overprojection in a thick section. The apparent size of dense particles or phase regions is enlarged so that the apparent volume fraction is increased.

and in many cases the only use of the distribution data is to report a mean and standard deviation. By combining the disector and point sampled intercepts, results can be obtained without these limitations.

Most of the examples described here have treated images as perfect two-dimensional representations of an ideal plane through the specimen, to reveal the internal three-dimensional structure. For an opaque sample examined by reflected light, or in the SEM or AFM, this is pretty much true (the depth to which electrons sample beneath the surface in the SEM is usually small compared to the lateral resolution), although in some cases (e.g., freeze fracture) it is legitimate to wonder if the plane being viewed is really representative or may be biased in position because of some weakness or other nonuniformity present in the structure.

For transmission imaging, either by light or electron microscopy, and for confocal light microscopy, the image is actually the projection of a finite depth, controlled either by the physical thickness of the slice or by the optical depth of the field of the instrument. If this thickness is small relative to the dimensions of the structures of interest, then the approximation of the image as an ideal 2D slice is acceptable, but if the thickness is great enough this assumption breaks down. Many of the simple relationships presented above have been adapted to the case of thick sections, although they generally become more complicated and interdependent, so that several parameters must be measured in conjunction.

For example, imaging a dense (dark) phase in a section of finite thickness produces a representation of that phase that is too large (overprojection) as shown in Figure 1.37. Determining the correct volume fraction of the phase requires measuring both its apparent area fraction and the surface area of the phase (S_V), because as shown in the sketch it is along the boundaries of the phase region that the slopes of the sides produce a greater apparent area than would be seen in a much thinner section. The relationship is

$$V_V = A_{A_{true}} = A_{A_{apparent}} - S_V \cdot \tfrac{t}{4} \qquad (1.12)$$

where t is the actual slice thickness. Another way to determine a correct volume fraction is to measure the area fraction in two images from slices of different (known) thicknesses, and calculate the true volume fraction from that. In all cases, it is necessary to accurately know the actual thickness of the imaged slice, which is not always easy to determine.

There are many other stereological relationships available. Many of them have subtle derivations that involve integral geometry and have been developed by experts.

But in most cases the application of the technique is straightforward, involving a specific procedure for sample preparation, with the use of an appropriate grid to achieve IUR probes, followed by counting or measurement of the intersection of the grid with the features of interest. In the chapters that follow, computer-based image processing methods will be used to perform the steps that accomplish the measurement procedures. In many cases this is more efficient than carrying out the same steps by hand, using photographs with grids (usually printed on transparent sheets to lay on top of the photos), with manual marking and counting of the hits. But the underlying principles and proper interpretation of the results for each procedure remain the same for entirely manual execution.

It is important to understand what each technique measures, and to select a method appropriate to the type of sample and features of interest. In addition to publications of results in applications-related journals, many stereological procedures are published in the *Journal of Microscopy* (Royal Microscopical Society, Oxford, U.K.) or in *Acta Stereologica* (published by the International Society for Stereology). Several recent books, *Unbiased Stereology* (C. V. Howard and M. G. Reed, Bios Scientific, 1998) and *Principles and Practice of Unbiased Stereology* (P. R. Mouton, Johns Hopkins University Press, 2002) cover primarily the new stereological techniques that emphasize design-based sampling procedures. *Practical Stereology, 2nd edition* (J. C. Russ and R. T. Dehoff, Plenum Press, 2002) covers both the classical and new methods, with many worked examples. A much older text, *Quantitative Microscopy* (R. T. Dehoff and F. N. Rhines, 1968, McGraw Hill), contains a good guide to classical stereology.

2 Image Acquisition

Having established in Chapter 1 that images provide a means by which to study the structure of food products, we must now consider how to acquire those images. Traditional photography is often sufficient if the measurements are to be performed manually, such as by placing transparent overlays containing grids onto photographic prints, and counting stereological events. But the emphasis in this text is on the use of computers to facilitate and in many cases automate those operations, so it is important to consider ways to digitize the images for storage in the computer.

One very important topic that is not addressed here at all is sample preparation. The range of techniques that are used is so broad for the various types of natural and prepared food materials that covering it would require many volumes and the expertise of many experienced researchers. In some cases nothing more than scattering particles on a contrasting substrate, or using a razor blade to cut a section is required. In other situations, rapid freezing followed by fracturing and coating is needed. Chemical fixatives and stains are widely used. Techniques for electron microscopy differ from those for light microscopy, and so forth. The current literature is generally the best source for reviews of existing techniques and reports of new ones. The goal is to preserve the fine details of structure without alteration or deterioration, with sufficient contrast to show that structure with the chosen imaging technology. The appropriate methods are generally the same for digital image processing as for conventional photographic recording.

SCANNERS

Photographic recording of images from macro cameras and microscopes is a well-known and widely practiced technique for obtaining images of food structure. Film is a very convenient way to record and store images, with well understood characteristics. Continuing to use film may be a good solution in many cases, as it allows direct comparison of images to existing archives that contain a rich treasure of images useful for various measurements. Relatively inexpensive scanners are available to digitize these images into a stored array of pixels within the computer.

Scanners are basically of two types. Flat-bed desktop scanners are primarily designed to scan prints using reflected light (although some units also have limited ability to handle transparencies and negatives). Since photographic prints have decidedly inferior dynamic range or gamut compared to the negatives from which they are produced, it is always preferable to scan the negatives rather than the prints if those are available. But flat bed scanners have another use that is less advertised, namely to directly image samples placed on them. Scattering particulates onto the

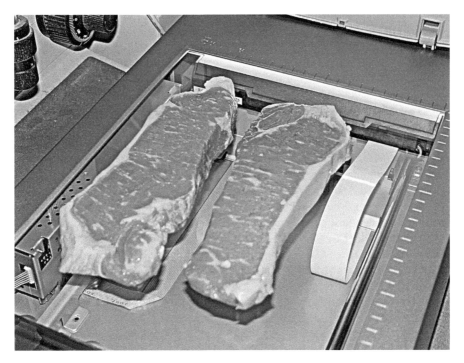

FIGURE 2.1 Using a flatbed scanner to directly image beef steaks.

scanner, or placing the cut surface of fruit, bread or meat onto the glass (as shown in Figure 2.1), provides a simple way to capture an image with quite high resolution and good color rendition. The sample must be relatively flat, because the depth of field is typically no more than about 1 cm. However, using cold foods may cause condensation on the underside of the glass.

The issues of spatial resolution and dynamic range come up again and again in discussing image acquisition, so it will be useful to deal with them here. Spatial resolution is usually defined in terms of points per inch (or dots per inch — dpi), and corresponds more-or-less to the size of structure that can be identified in the image (technically, the features must be at least twice the size of the measured points in order to be distinguished from their neighbors). Local contrast also plays a role in the ability to see small features, as well as be seen later on. Many desktop scanners currently offer optical resolutions of 1000 to 3000 dpi. Ignore any specification for interpolated resolution, as that simply expands the real data to produce a larger array of stored pixels but, like any other empty photographic magnification, does not reveal any additional information.

The ability to acquire an image with (for example) 2000 dpi of real resolution over a distance of 8 inches produces a huge amount of information. That would correspond to a 16,000 × 16,000 pixel stored image, or 256 million pixels, with each pixel representing a dimension of 12.5 μm on the specimen. This generates a large image file, but one that most modern desktop computers and software can accommodate (as an example, Adobe Photoshop® accepts images with dimensions up to

30,000 pixels on a side, and uses virtual memory on disk to handle images larger than will fit into memory). A digital camera connected to a macro lens can certainly achieve a spatial resolution of 2000 dpi, but over a much smaller distance of an inch or two. To image such a large area would require a corresponding reduction in spatial resolution. So for situations in which images of sections through food products can be obtained by cutting them and placing them on a flatbed scanner, doing so provides an extremely high quality result at a very low cost and with great convenience. Figure 2.2 shows a representative application.

With a scanner, there is no concern about focusing (provided the sample is relatively flat) or achieving uniform lighting. The hardware takes care of that (although avoiding the very edge of the platform, where light intensity may fall off slightly, is a good idea). There may be some clean-up problems, but these are at most a minor nuisance. The time required for scanning is typically tens of seconds, but that is usually fast enough. Scanners come with a variety of standard interfaces, ranging from older SCSI (small computer systems interface) to the newer, faster, and more trouble free USB (universal serial bus) or IEEE 1394 (widely known as firewire). Many scanners are bundled with software such as Adobe Photoshop that provides a good platform for image acquisition, storage, printing, and some of the processing techniques discussed below and in subsequent chapters.

Scanners also typically have a high dynamic range. This is usually specified as bits of information in each channel. The scanners have sensors that read the red, green, and blue light intensity separately. This is typically done by using one or several linear sensors with colored filters that are mechanically scanned across the area. The maximum dynamic range is determined in part by the well size of the detectors, that is, the number of electrons that can be accommodated, and the noise level of the electronics used for readout. The readout is much slower than a typical digital camera (and considerably slower than a video camera). This results in less noise and a higher dynamic range.

In any case, the individual red, green, and blue light intensities from points on the sample are digitized — converted to a numerical value — for transmission to and storage in the computer. An 8-bit image, for example, uses one computer byte to store each of the RGB channel values, and since 2^8 is 256, it can represent 256 discrete brightness values in each channel. That may seem like a lot, and for some purposes it is. Display of the image on the computer screen for human viewing, and printing the image out as hardcopy, does not require more than 8 bits and can be adequately accomplished in most cases with less. But for many of the image processing operations to be described, and to detect small variations in brightness that may represent local structural details in an image that has a large overall contrast range between bright and dark values, it is best to have more tonal resolution than the 256 values provided by 8 bits. Indeed, many professional photographers suggest that all work on digital photographs be performed in a 16 bit space, only reducing the image to 8 bits for the final printout (because printers cannot handle and do not need the increased amount of information).

Actually, few detectors and scenes have 16 bits of information (2^{16} = 65536). Photographic prints do not typically have as much as 8 bits of data, although the negatives can be much better as discussed below. In astronomy, where stars are very

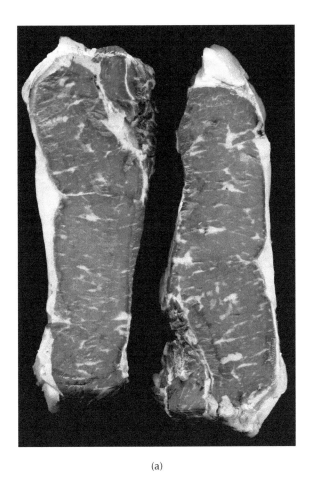

(a)

FIGURE 2.2 Image of beef steaks from Figure 2.1. Thresholding of the fat (b) allows measurement of the total volume fraction of fat, and the size, shape, and spatial distribution of the individual areas of fat in order to quantify the marbling of the cut.

bright and space is very dark, cameras cooled to liquid helium temperatures record such a range of brightness data, but it requires very specialized detectors. The actual performance of the detectors in most flat bed scanners is more like 12 bits (2^{12} = 4096 values), which is still a great improvement over a simple 8-bit image and plenty for most of the applications used. Because of the organization of computer memory into bytes, each of 8 bits, it is usually convenient to store any image with more than 256 values in two bytes per channel, and hence these are usually called "16-bit images." In such an image, the measured values are multiplied up to fill the full 16-bit range of values, with the result that very small differences in value (for instance, the difference between a brightness value of 10,000 and 10,005) are not significant.

A good way to understand the importance of high bit depth for images is to consider trying to record elevations of the Earth's surface. In the Himalayas, altitudes reach over 25,000 feet, so if we use 8 bits (256 discrete values), each one will

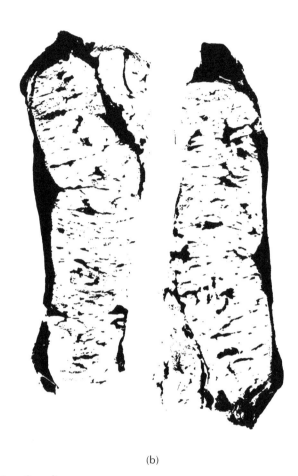

(b)

FIGURE 2.2 (continued)

represent about 100 feet, and smaller variations cannot be recorded. With that resolution for the elevation data, the entire Florida peninsula would be indistinguishable from the ocean, and much fine detail would be lost everywhere. Using 12 bits (4096 values) allows distinguishing elevation differences of slightly over 6 feet, and now even highway overpasses are detectable. If 16 bit data could be acquired, the vertical resolution would be about 5 inches. With only 8 bit images, a great amount of information is lost, although it typically requires image processing to make it visible since human vision does not detect very slight differences in brightness.

In this text, because images will come from many different sources, some of which are 8 bit, some 16 bit, and some other values, the convention will be adopted that a value of 0 is completely dark, a value of 255 is completely bright, and if the available precision is more than 8 bits it will be represented by a decimal fraction. In other words, a point on a feature might have a set of brightness values of R = 158.53, G = 74.125, B = 146.88 (if imaged with an 8 bit device, that would have been recorded as (R = 158, G = 74, B = 146). Figure 2.3 illustrates the ability

FIGURE 2.3 Reading the color values at a point in an image (the original color image is shown as Color Figure 2.3; see color insert following page 150). The point measured is on a neon purple eggplant; the image fragment is enlarged to show individual pixels.

of most programs to read out the color values. How such values correspond to perceived colors is discussed later in this chapter.

Most flat bed scanners provide 16 bit per channel data, although the software may truncate these to a lower 8-bit precision unless the user elects to save the full data (file sizes increase, but storage is constantly becoming faster and less expensive). Many manufacturers describe scanners that store images with 8 bits in each of the three color channels as 24-bit RGB scanners, and ones that produce images that occupy 16 bits per channel as 48-bit RGB scanners, even though the actual dynamic range or tonal resolution of the image is somewhat less.

Some flatbed scanners provide a second light source (or sometimes a mirror) in the lid and can be used to scan transparencies, but much better results are obtained with dedicated film scanners. Mechanically, these are similar to but smaller than the flatbed scanners. Optically they provide very high spatial resolution, typically in the 3500 to 5000 dpi range, which corresponds to the resolution that film can record. Most units have built-in autofocusing and color calibration, and require no particular skill on the part of the operator. Some have special software, either built into the unit or provided for the host computer, that can selectively remove scratches or other defects present on the film. Because these can also remove real detail in some cases, it is usually best to examine the raw data before allowing the software to "enhance" the result. Most film scanners come with software that includes color calibration curves for both positive (slide film) and negative (print film) color films, and give excellent results.

An inexpensive negative scanner combined with an existing film camera back used with macro lenses or a microscope represents an extremely cost-effective way to accomplish digital imaging. Typical 35 mm color slide film has a spatial resolution of about 4500 × 3000 points and a tonal resolution of one part in 4000 (12 bits per RGB channel). Color negative film (used for prints) has less dynamic range. All of this can be captured by the scanner, with scan times of tens of seconds per image. No affordable digital camera produces images with such high spatial or tonal resolution. But the drawback of the film-and-scanner approach is the need to take the entire roll of pictures and develop them before knowing if you have a good photo, and before you can perform any measurements on the image. It is the immediacy of digital cameras rather than any technical superiority that has encouraged their widespread adoption.

DIGITAL CAMERAS

Digital still cameras (as distinct from video cameras, which use similar technology but have much less resolution and much more noise in the images) have taken over a significant segment of the scientific imaging market. Driven largely by a much greater consumer marketplace for instant photography, and spurred by the Internet, these cameras continue to drop in price while improving in technical specifications. But it is not usually very satisfactory to try to adapt a consumer digital camera to a scientific imaging task. For these applications, a camera back that can accept different lenses, or at least attachments that will connect to other optical devices such as a microscope, is usually more appropriate. (The other major issue with consumer cameras is the data storage format, which is discussed below.)

The same technical issues that arise for the scanner also apply to cameras. We would ideally like to have a high spatial resolution along with a high dynamic range and low noise. High spatial resolution means a lot of detectors in the array. At this writing, a typical high-end consumer or low-end professional camera has about 5 to 8 million "pixels" and a few professional units have three times that many. A note of caution here — pixel has several different meanings and applications. Some manufacturers specify the number of pixels in the stored image as the measure of resolution, but extract that image by interpolating the signals from a smaller number of actual detectors on the chip. Many use the word pixel for the number of diodes that sense light, but choose to ignore the fact that only one fourth of them may be filtered to receive red light and one fourth for blue light, meaning that the color data is interpolated and the actual resolution is about half the value expected based on the pixel count.

Since most digital cameras are intended for, and capable of recording color images, it is worth considering how this may be accomplished. It is perfectly possible, for example, to have a camera containing a single, high resolution array of detectors that respond to all of the visible wavelengths. Actually, silicon-based detectors are quite inefficient at detecting blue light and have sensitivity well into the infrared, but a good infrared-cutoff filter should be installed in the optical path to prevent out-of-focus infrared light from reaching the detector and creating a fuzzy, low contrast background. If a complete camera system is purchased, either one with a fixed lens or one from the same manufacturer as the microscope, this filter is usually included. But if you are assembling components from several sources, it is quite easy to overlook something as simple as this filter, and to find the results very disappointing with blurry, low-contrast pictures.

If a single detector array is used, it is then necessary to acquire three exposures through different filters to capture a color image (Figure 2.4a). A few cameras have used this strategy, combining the three images electronically to produce a color picture. The advantages are high resolution at modest cost, and the ability to achieve color balance by varying the exposure through each filter. The penalty is that the time required to obtain the full-color picture can be many seconds, the filters must be changed (either manually or automatically), and during this long time it is necessary for the specimen to remain perfectly still. Also, vibration or other interference can further degrade images with long exposure times. And of course, there is no live color preview with such an arrangement.

At the other extreme, it is possible to use three chips with separate detector arrays, and to split the incoming light with prisms so that the red, green, and blue portions of the image fall onto different chips (Figure 2.4b). Combining the signals electronically produces a full-color image. This method is expensive because of the cost of the three detectors, electronics, prisms, and alignment hardware. In addition, the cameras tend to be fragile. The optics absorb much of the incoming light, so brightly lit scenes are needed. Also, because of the prisms, the satisfactory use of the three-chip approach is usually limited to telephoto lenses. Short focal length lenses direct the light through the prisms at different angles resulting in images with color gradients from top to bottom and/or left to right. Three-chip cameras are used for many high-end video cameras, but rarely for digital still cameras.

Many experimental approaches are being tried. The Foveon® detector uses a single chip with three transistors stacked in depth at each pixel location (Figure 2.4d). The blue light penetrates silicon the least and is detected near the surface. Green and red penetrate farther before absorption, and are measured by transistors deeper beneath the surface. At present, because these devices are fabricated using complementary metal oxide on silicon (CMOS) technology, the cameras are fairly noisy compared to high performance charge coupled device (CCD) cameras, and combining the three signals to get accurately calibrated color remains a challenge.

The overwhelming majority of color digital still cameras being used for technical applications employ a single array of transistors on a CCD chip, with a filter array that allows some detectors to see red, some green, and some blue (there are a few consumer cameras that use other combinations of color filters). Various filter arrangements are used but the Bayer pattern shown in Figure 2.4c is the most common. It devotes half of the transistors to green sensitivity, and one quarter each to red and blue, emulating human vision which is most sensitive in the green portion of the spectrum.

With this arrangement, it is necessary to interpolate to estimate the amount of red light that fell where there was no red detector, and so on. That interpolation reduces resolution to about 60% of the value that might be expected based on the

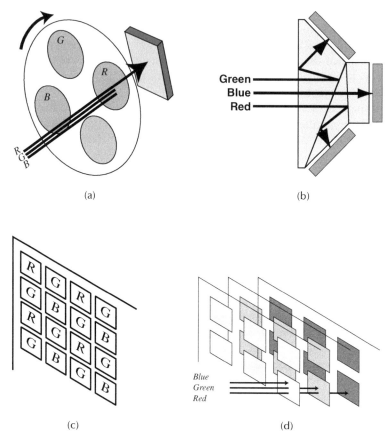

FIGURE 2.4 Several ways for a digital camera to acquire a color image: (a) sequential images through colored filters with a single chip; (b) prisms to direct colors to three chips; (c) Bayer filter pattern applied to detectors on a single chip; (d) Foveon chip with stacked transistors.

number of camera transistors, and also introduces problems of incorrect colors at boundaries (e.g., is the average between a red pixel and a green pixel yellow or grey?). Interpolation is a complicated procedure that each manufacturer has solved in different (and patented) ways. Problems, if present, will often show up as zipper marks along high contrast edges. The user typically has no control over this except to not expect to measure features with a width of only one or two pixels in the image. In the measurement chapters we will typically try to limit measurement to features that cover multiple pixels.

The principal reason for wanting a high number of pixels in a camera detector is for more than dealing with small features. Usually that can be done best by applying the appropriate optical magnification, either with a macro lens or microscope objective. If all of the features of interest are the same size, then even a fairly small image (at current technology levels that is probably a million pixels, say 1200 × 900) is quite useful provided the image magnification is adjusted so the features are large enough to adequately define size and shape.

Difficulties arise when the features cover a range of sizes. Large features have a greater probability of intersecting the edge of the image frame, in which case they cannot be measured. In the example just given, if the smallest features of importance are 20 pixels wide, and the largest features are 20% of the image with (say about 200 pixels) then it is possible to satisfactorily image features with a 10:1 size range. If the features present actually cover a greater range of sizes, multiple images at different magnifications are required to adequately image them for measurement. But sometimes we need to measure spatial arrangements, such as how the small features may be clustered around (or away from) the large ones. In that case there is really no good alternative to having an image with enough pixels to show both the small features with enough resolution and the large ones with enough space.

Figure 2.5 illustrates this problem. The fragment of the original image, captured with a high resolution digital camera, shows both large and small oil droplets and also details such as the thin layer of emulsifier around the periphery of most droplets. Decreasing the pixel resolution by a factor of 6, which is roughly equivalent to the difference between a 3 megapixel digital camera and the resolution of a video camera, hides much of the important detail and even limits the measurement of droplet size.

Despite their poor resolution and image noise, video cameras are sometimes used as microscope attachments. Historically this was because they existed and (driven by a consumer marketplace) were comparatively inexpensive, produced a real-time viewable image, and could be connected to "framegrabber" boards for digitization of the image. Currently, the only reason to use one is for those few applications that require image capture at a rate of about 30 frames per second. Most studies do not. Either the specimens are quite stable, permitting longer exposures, or some dynamic process is being studied which may require much higher speed imaging. Specially designed cameras with hard disk storage can acquire thousands of images per second for such purposes.

Human vision typically covers a size range of 1000:1 (we can see meter-sized objects at the same time and in the same scene as millimeter-sized ones). A 4×5 photographic negative can record images with a satisfactory representation of objects over a 150:1 size range. But even a high performance digital camera with 6 million pixel resolution (reduced somewhat by interpolation of the color information) is limited to about 20:1 in size range. That may be frustrating for the human observer who sees more information in the microscope eyepiece, and is accustomed to seeing a wider range of object sizes in typical scenes, than the camera can record. This topic arises again in the context of feature measurement.

The number of camera pixels is most significant for low magnification imaging. At high magnification, the limit to resolution in the image is typically the optical performance of the microscope (and perhaps the specimen itself and its preparation). At low magnification, where a large field of view is recorded, an image with a large number of pixels provides the ability to resolve small features and provide enough pixels to define their size and shape.

In addition to the resolution or number of pixels in the camera, the same issues of well size, noise level and bit depth arise as for scanners. Many consumer cameras, particularly low-cost ones such as are appearing in telephones and toys, use CMOS

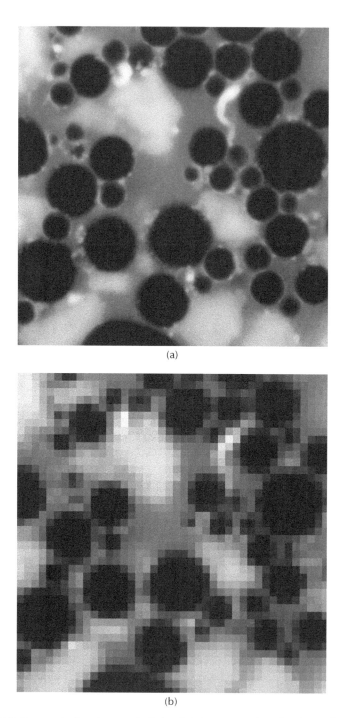

(a)

(b)

FIGURE 2.5 Fragment of an image of oil droplets in mayonnaise, showing a size range of oil droplets and a thin layer of emulsifier around many: (a) high resolution camera; (b) effect of a 6× decrease in pixel resolution, losing the smaller details.

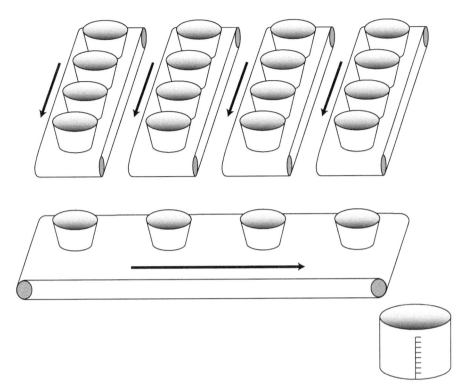

FIGURE 2.6 Schematic diagram of the functioning of a CCD detector. The individual detectors function like water buckets to collect electrons produced by incoming photons. Shifting these downwards in columns dumps the charge into a readout row. Shifting that row sideways dumps the charge into a measuring circuit. The overall result is to read out the image in a raster fashion, one line at a time.

chips. These have the advantage of low cost because they can be fabricated on large wafers and can include much of the processing electronics right on the chip, whereas the CCD devices (Figure 2.6) used in higher-end scientific cameras are more expensive to fabricate on smaller wafers, and require separate electronics. CMOS detectors are also being used in a few prosumer (high-end consumer/low-end professional) cameras, because it is possible to combine several discrete chips to make one large detector array, the same size as a traditional 35mm film negative, with about 15 million pixels.

But the CMOS cameras have many drawbacks, including higher amounts of random noise and fixed pattern noise (a permanent pattern of detector variability across the image) that limits their usefulness. Also, the extra electronics on the chip take up space from the light-sensing transistors and reduce sensitivity for low-light situations. This can be overcome to some extent by using microlenses to collect light into the detector, but that further increases the point-to-point variations in sensitivity and fixed pattern noise. It is likely that CCD detectors will remain the preferred devices for scientific imaging for the foreseeable future.

Even for CCD chips, the drive for lower cost leads manufacturers to reduce the chip size (at the same time as increasing the number of transistors) in order to get more devices from a given wafer. That makes the individual transistors smaller, which reduces light sensitivity and well size. With some of the current small chips having overall dimensions only a few mm (a nominal one third inch chip has an actual active area of 3.6×4.8 mm, while a one quarter inch chip is only 2.4×3.2 mm), the individual transistors are so small that a maximum signal to noise ratio of 200:1 can barely be achieved. That is still enough for a video camera, which is only expected to produce a dynamic range between 50:1 and 100:1, but not enough for a good digital still camera.

Most digital still cameras record images with at least 8 bits (256 brightness values) in each RGB channel. Internally, most of them have a somewhat greater dynamic range, perhaps 10 bits (1000 brightness values), which the programs in their internal computers convert to the best 8 bits for output. With many of the cameras it is also possible to access the full internal data (RAW data) to perform your own conversion if desired, but except in cases with unusual illumination or other problems the results are not likely to produce more information than the built-in firmware. Also, the raw internal data is usually linear but for most cameras the output is made logarithmic, to achieve a more film-like response, and this reduces the effective bit depth.

Obtaining more than 8 bits of useful data from a digital camera usually requires cooling the chip to reduce the noise associated with the readout process. Extreme cooling as for astronomical imaging is not needed. Peltier cooling of the chip by about 40 degrees, combined with a somewhat slower readout of the data, is typically enough to produce images with 10 or even 12 bits (approximately 1000 to 4000 brightness levels) in each channel. Cooling the chip also helps to reduce thermal noise in the chip itself when exposure times are longer than about one-quarter to one-half second. Cooled cameras are often used for microscopes when dark field or fluorescence imaging is needed, whereas 8 bits is probably enough when only bright field images are captured.

SCANNING MICROSCOPES

Several types of microscopes produce images by raster scanning and are capable of directly capturing images for computer storage (Figure 2.7). This includes the scanning electron microscope, the scanning confocal laser microscope, and the atomic force microscope, as well as other similar scanned probe instruments. These use very different physical principles to generate the raster scan and the imaging signal. The confocal microscope uses mirrors to deflect the laser beam, while the SEM uses magnetic fields to deflect an electron beam, the AFM uses piezoelectric devices to shift the specimen under a fine-pointed stylus. But they share many of the same attributes in so far as the resulting image quality is concerned.

The desire for spatial resolution and dynamic range are the same as for any other image source, with the single exception that these devices produce monochrome rather than color images. Confocal microscope images are often displayed as color,

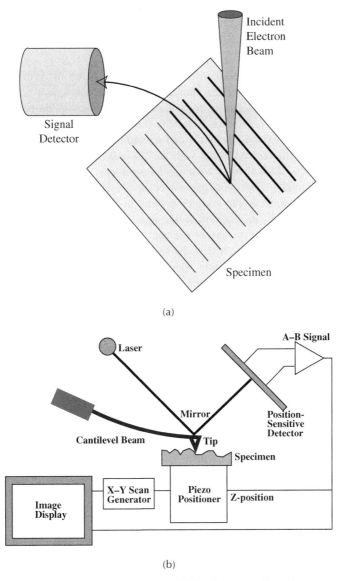

FIGURE 2.7 Schematic diagram of image acquisition from scanning microscope: (a) in the SEM the beam follows a raster pattern across the specimen and various signals such as secondary electrons or X-rays emitted from the surface are collected; (b) in the AFM the specimen is moved by actuators beneath a fixed or oscillating tip to record surface elevation and the various tip-surface interactions.

but that is accomplished by capturing three images with different laser wavelengths and combining them afterwards. AFM images are frequently displayed using color, but the original data are monochromatic, and are rendered in pseudocolor as a graphics visualization aid. When multiple signals, such as elevation, lateral force, and the like, are displayed in different color channels, they are combined after being acquired separately. The same thing is true of the SEM, in which multiple signals such as secondary electrons, various elemental X-ray intensities, and so on, are collected by different detectors and digitized separately.

The pixel size of the acquired images is controlled by the scanning circuitry, and is usually determined by the manufacturer of the instrument according to the actual resolution of the signals, or at least to whichever of the several recorded signals offers the best resolution. Typically this results in an image of modest size, rarely more than 1000×1000 pixels and sometimes much less. Historically many of these instruments had problems with the linearity of their scans, and the image distortions that could result, but most modern instruments have solved these difficulties.

The use of analog displays to present the image data, which requires digitization of the signal with an analog to digital converter in the attached computer, is increasingly being replaced by digitization of the signal in the microscope instrument itself, with internal storage and display of the captured image on a computer monitor. In that case, it is usually easy to access the saved image file for subsequent processing and measurement, although some translation from private file formats may be required. This also usually solves the problem of recording square pixels, meaning that the pixel spacing is the same in the X and Y directions. With analog systems and separate digitizers that was often hard to accomplish (as indeed it can be with analog video cameras attached to light microscopes).

With all of these scanning microscopes the key parameter that controls image quality is the scan speed. With a fast scan speed, capable of producing a frequent update on the screen so that specimen positioning, focusing, and other manipulations can be comfortably performed, the image is likely to appear quite noisy or to be smeared out horizontally. Practically all scanning instruments use the horizontal direction as the fast scan direction in the raster and the vertical direction for the slow scan.

With the SEM, the typical cause of the noisy image arises from the use of a very small beam current and/or low voltages (in order to achieve small beam diameter, shallow penetration into the sample, and good spatial resolution, while reducing charge buildup), which in turn produces a very small signal. The grainy random noise or speckle in the image is dominated by the statistical variations in the output signal even from a uniform specimen surface, and can hide important details. Slowing the scan rate down as shown in Figure 2.8 allows more signal to be generated at each location, requires less amplification of the signal, reduces the random fluctuations, and produces a more noise-free image.

Too slow a scan rate not only makes it difficult to position the specimen and to focus, it also allows charge from the beam to accumulate on parts of the image. This in turn produces a variety of effects, from very bright and dark regions where the charge has altered the emission or collection of electrons, to image distortions where the charge deflects the electron beam. Charging can be minimized by careful coating of the specimen with a thin conductive layer and the use of low beam voltages, but very slow scan rates increase the problem. Many foods and food products are not naturally electrically conducting.

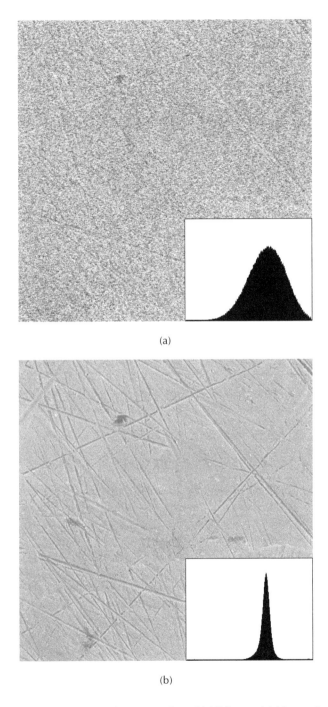

(a)

(b)

FIGURE 2.8 SEM images of scratches on a surface: (a) 1/30 second (video rate) scan showing random noise; (b) 10 second slow scan showing reduced noise and ability to see fine details. The histograms show the variation in grey scale values for the pixels in each image.

In the AFM, the effect of fast scan rates is to cause the stylus to skip over details on the surface, and generally fail to follow changes in slope or abrupt depressions. Slow scans capture the topological details of the surface but can cause line-to-line variations due to shifts in signal amplification. In all cases, adjusting the amplifier time constants to match the scan rate is very important to prevent smearing or blurring the image in the fast scan direction.

When the confocal scanning laser microscope is used for fluorescence microscopy, the effect of a fast scan rate is similar to the situation in the SEM, because the number of generated photons is small and the statistical fluctuations large. But scanning slowly may cause bleaching that reduces the fluorescence intensity. It may be preferable to scan relatively quickly but to add together multiple scans to accumulate more intensity.

With all of these scanning microscopes, the noise characteristics and actual spatial resolution of signals from the image may be quite different in the slow- and fast-scan directions, which can pose challenges for image processing that will be discussed below and in the next chapter. The dynamic range of the signals varies widely. The SEM typically needs no more than 8 bit images to capture all of the significant information, although there may be some cases in which the X-ray signals require 16 bits. Eight bits is also enough for bright field confocal microscope images but not for fluorescence microscopy, which can have very large brightness differences between the dark and bright areas and requires 12 bits or more. The elevation values measured by the AFM can easily cover a range of up to 100 μm with a resolution of a few nanometers, requiring at least 16 bits and sometimes more. Many scanned stylus instruments record 32 bit values (in terms of the mapping of the earth example used previously, that would produce vertical resolution of less than 1/10000th of an inch while covering the full height of Mount Everest).

FILE FORMATS

Image files can be quite large. The example cited above of a scanned image with 2000 dpi pixel spacing covering 8 inches on a side, in RGB color with 16 bits per channel, would produce a file of 1.5 gigabytes. Even a more routine 2000×2000 pixel RGB color 8 bit per channel image produces a 12 megabyte file. Storing image files on the computer's primary hard disk fills it up quickly. Storing them on a medium like writable CDs or DVDs is inexpensive and open-ended in terms of capacity, but too slow for efficient processing or measurement. And transmission of large image files over the internet can be painfully slow.

Image compression seeks to produce smaller file sizes, to speed file transmission and reduce storage requirements, particularly in digital cameras. All consumer cameras, many high-end, and some professional cameras record images in a JPEG (Joint Photographers Expert Group) format that reduces file sizes by factors of 5 to as much as 50 times. That is especially important for the camera manufacturers, since the cost of the memory comprises a significant part of the cost of the camera. It also speeds up the overall camera operation because in most cases compressing the image and storing a smaller file is actually quicker than storing the larger original file.

JPEG compression carries with it a very high price in terms of the image contents. The goal of all compression techniques is to preserve those details and characteristics of the original image that are noticed by human vision and which convey information to the viewer necessary for the recognition of familiar objects. The original JPEG technique reduces the amount of color information, applies a discrete cosine transform to 8×8 pixel blocks in the image, rounds off the resulting coefficients to a set of quantized values (and eliminates small terms, particularly for color data) and then encodes the result. The technique has been widely used because the algorithm can be embedded into chips as well as carried out in software, and it is quite fast.

The JPEG method, however, discards information from the image and, in addition to the appearance of visually distracting blockiness, alters colors, reduces color resolution, shifts edges, removes real texture and inserts artificial texture (lossy compression). All of these problems are minor if the purpose is to allow recognition of pictures of the kids in the backyard, or a reminder of a visit to the Eiffel tower, but for scientific imaging it creates serious problems. First, we are usually not trying to simply recognize familiar structures, but to detect or characterize unfamiliar things. Second, we may want to measure those structures, and alterations in position, dimension, shape and color are not acceptable.

There are other compression methods available as well. Wavelet compression, which is part of the newer JPEG2000 standard, uses a wavelet transform instead of the discrete cosine method. The results do not show the visible blockiness that the original JPEG method produces (because of its use of 8×8 pixel blocks), but it has the unfortunate characteristic of discarding more detail in some parts of the image (where there was a lot of original detail present) while keeping it elsewhere. Fractal compression, on the other hand, inserts artificial detail into the image everywhere. This prevents enlargements from appearing to be interpolated (empty magnification), but the detail is not real, just pixel patterns borrowed from elsewhere in the image. Real details are replaced by visually plausible borrowings. Fractal compression is very asymmetric, meaning that it takes much longer to perform the compression than to decompress the stored image.

One of the best ways to see what lossy compression discards from an image is to perform the compression on an original image and then subtract the result from the original. As shown in Figure 2.9, the differences reveal the changes in pixel brightness (and also color), displacement of edges, alteration of texture, and so forth. Detecting whether an image has ever been subjected to lossy compression can usually be done by looking just at the color information. A discussion of the various color spaces to which images can be converted is presented below. In HSI space the hue channel, or in L-a-b space the a and b channels, reveal the loss of resolution by their blurry or blocky appearance as shown in Figure 2.10.

The safest recommendation for scientific imaging is to avoid any lossy compression entirely. Most high-end cameras, and all scanning microscopes with digital output, provide lossless image formats. There may still be a small amount of compression possible, by finding repetitive sequences of values in different parts of the image and representing them with an abbreviated code. The LZW (Lempel–Ziv–Welch)

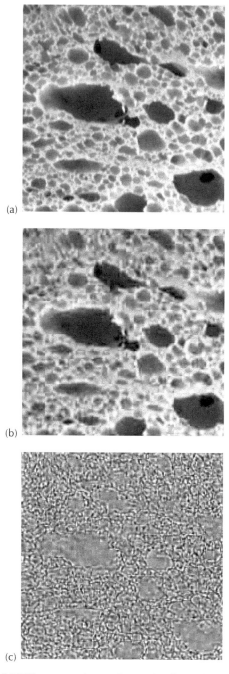

(a)

(b)

(c)

FIGURE 2.9 Effect of JPEG compression on image detail: (a) fragment of high-resolution image of sliced bread; (b) same image after JPEG compression reduced the file size by 25×; (c) difference between original and compressed images. Note the 8 × 8 pixel blockiness, shifting of edges, alteration of grey scales, and loss of texture and detail due to the compression.

FIGURE 2.10 The hue channel from the market image shown in Color Figure 2.3 (see color insert following page 150): (a) original; (b) after JPEG compression.

method is one of the more widely used techniques for achieving this lossless compression, and it is built into most computer programs routines for reading and writing standard formats such as TIFF (tagged image file format). But LZW compression does not produce a very great compression, rarely as much as 50% (as compared to the

JPEG factor of 30 or more) and so it does not really help much with storage or transmission of large image files.

Certainly the most widely-used lossless image format, readable by most image processing programs on Windows, Mac, and Unix/Linux computers, is the TIFF format first proposed by Aldus, now sponsored by Adobe, and supported by open-source library code downloadable from the Internet. TIFF files (type *.tif under Windows) can accommodate a variety of bit depths, in either monochrome or RGB color, and can have multiple layers, which is useful for comparing or combining multiple images. TIFF is probably the safest choice for storing images that you may want other people or other programs to be able to read.

There are certainly other choices possible, with dozens of formats used in various contexts and programs. The PSD file format, introduced by Adobe Photoshop and now used by some other imaging programs, does everything TIFF does and also allows layers containing text or annotations, or image adjustments such as alterations in contrast, color or brightness, to be saved without altering the original image. The PNG format is gaining increased acceptance particularly for Web applications because it is public domain, much simpler than TIFF (with fewer variants), but is not as widely supported. Many cameras and scanning microscopes have their own internal storage format that is proprietary. Fortunately, most devices can save the image in a standard format, most commonly TIFF.

The user probably needs to know little about the internals of the file format chosen beyond making sure that it is truly lossless, readable by whatever other programs or people will need access, and that it saves the full range of image data. In some cases, images that appear to be saved as lossless TIFF files by scanning microscopes are actually reduced to a smaller bit depth and/or converted to indexed color with just a small number (usually 256) of color values that approximate the original colors well enough to show the image on the computer monitor. Usually described as "indexed color," this must be avoided if the goal is to actually use the images for anything later on.

It can be argued that the measurement processes introduced in Chapter 1 and elaborated in following chapters also constitutes an effective compression of the image. Typical image measurement data comprise no more than a few dozen numbers (for instance a size distribution) and sometimes only a single bit (whether or not a particular type of feature or defect is present), as compared to the enormous size of the original image file. But in general, storage of the images themselves is important even in quality control situations because it allows re-examination of the original images if new questions arise later on.

COLOR ADJUSTMENT

Digital images with red, green, and blue channels acquired from cameras or scanners that use colored filters to select ranges of wavelengths for each sensor cannot measure true color. The number of combinations of different wavelengths and intensities of light that would produce the same output from the detector is practically infinite. So if measuring the actual color of a food or food product is important — and it often is — only a spectrophotometer will provide the needed information.

In practical terms, however, it is often possible to assure that the visual presentation of color from a subject matches that from a standard. The rather common requirement of assuring that the appearance of color in a printed image matches the real thing is routinely met in the advertising and catalog industry. Every time someone makes a catalog purchase, he or she expects the delivered item, such as an article of clothing, to match the picture in the catalog. Accomplishing this isn't trivial, but it is routine. And it is no less of a headache for the traditional film photographer than it is for the digital photographer.

In both cases, the problem is that the recorded color in the image is the result of the color of the subject (how it absorbs and reflects light) combined with the color and placement of the light source(s), and the response of the detector (whether chip or film). The universally accepted way to deal with this is to take a picture of a known color target with the same camera and under the same lighting conditions. Sometimes this is simply the first image on the roll of film, and sometimes because the lighting or geometry is hard to maintain constant, it is something that can be included in every exposure.

The simplest target is just a neutral grey card, or at least some region of the image that was a medium grey without any color. A standard 18% reflectance grey card used by photographers can be used, but is not essential. In the resulting image, the red, green, and blue intensities in that region can be measured and should be equal (equal RGB intensities produce a colorless grey result). If the region has a discernible color shift, then adjusting the gain of the channels can make them equal and restore neutral grey to the test region, and hopefully balance the color intensities in other regions. That is simple, and often automated in software programs such as Adobe Photoshop, but not quite flexible enough in many cases.

Much more can be accomplished if, in addition to the grey region, there are known black and white areas as well. These should also be colorless, and should correspond to RGB values near zero and 255 (black and maximum) respectively but without clipping (being darker or brighter than the actual measurement range of the camera). With the end points for each color channel established, and one neutral point in the middle, it is possible to establish curves that adjust the values in each channel to produce visually realistic and balanced colors in many cases. Figure 2.11 shows an example.

For greater accuracy in color adjustment it is necessary to take into account the fact that the color filters used in digital cameras (and for that matter the absorption of light in the color-sensitive dyes of traditional films) cover a range of wavelengths, and that (just as in the human eye, as illustrated in Figure 2.12) these overlap somewhat. Photographing a calibration target with known red, green, and blue patches will produce regions in the image that can be measured to determine the intensities in the RGB channels. Ideally, they should consist of just one color. The red area should have $R = 255$, $G = 0$, $B = 0$, or close to it, and similar values should be found in the green and blue regions.

The typical result is that the regions have significant intensities in the other channels as well, resulting from the interaction of the light source with the subject and the width of the color filters in the camera. Mathematically, the problem can be solved by setting up a 3×3 matrix of values, with the fraction of maximum intensity

(a)

(b)

FIGURE 2.11 Adjusting color using neutral tones (Color Figures 2.11(a) and (b)); see color insert following page 150): (a) original; (b) adjusted; (c) curves for each color channel in which the horizontal axis is the original value and the vertical axis is the corrected value. The three reference points (white in the flag stripe, grey on the flagpole, and black in the tree shadows) are shown on the plots.

in each channel measured in each of the target regions. Inverting this matrix produces a set of tristimulus correction values that can be used to correct the entire image. Multiplying the correction matrix by the measured RGB intensities for each pixel produces a new set of corrected RGB values that replace the originals. The values

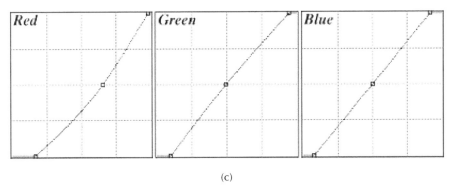

(c)

FIGURE 2.11 (continued)

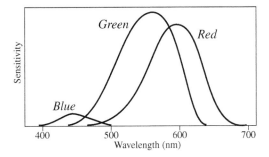

FIGURE 2.12 Relative sensitivities and wavelength coverage of the three types of cones in the human eye.

off the matrix diagonal are usually negative, indicating the need to subtract some of the response from other channels. The result is to make the colors in the target regions exactly red, green, and blue, and to adjust the colors of other features in the image as well. Figure 2.13 shows a practical example in the quality control inspection of a cooked pizza.

When this correction is large, the resulting image may still have some slight color cast in the neutral grey regions. This arises because color space (about which more below) is not quite uniform. Straight lines in most coordinate systems used to map color space do not mix colors exactly or preserve hues. There is a color space (CIE, for la Commission Internationale de l'Eclairage) that does behave in this ideal way, which is why it is used for applications such as broadcast television, but digital cameras and most image processing software work in simpler coordinate systems and accept slight nonlinearities. The combination of the neutral grey balancing step following tristimulus correction usually produces satisfactory results.

Although this procedure is not very complicated, it does require the use of a color standard. For macro photography the process is fairly simple, and color charts like the one shown are available from companies such as GretagMacbeth who base their entire business on assuring that color photographs (film or digital) come out right, and that the appearance of the image on paper and on the computer monitor

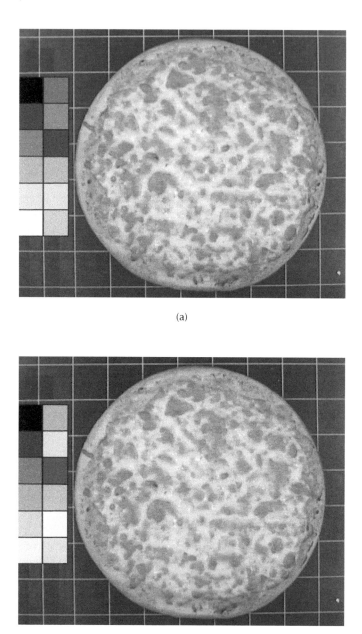

(a)

(b)

FIGURE 2.13 (See color insert following page 150.) Adjusting color with tristimulus coefficients: (a) original image including color reference chart; (b) corrected result using tristimulus coefficients. The table shows the measured intensities in the red, green, and blue regions of the color chart, their division by the maximum value of 255, and the inverse matrix with the coefficients used to generate the corrected result when applied to the entire image.

Area	Measured Intensities		
	Red	Green	Blue
Red	132.95	63.34	61.42
Green	63.15	133.56	64.03
Blue	56.54	55.71	128.86

Normalized Intensity Matrix (Divided by 255)

0.52137255	0.24839216	0.24086275
0.24764706	0.52376471	0.25109804
0.22172549	0.21847059	0.50533333

Inverse Matrix (Tristimulus Coefficients)

Result	Channel		
	Red	Green	Blue
Red	2.717942	−0.9443239	−0.8262528
Green	−0.8998933	2.72109353	−0.9231738
Blue	−0.8035029	−0.7620677	2.7405428

is the same. Color standards are harder to use in microscopy, and also more urgently needed there because the color temperature of the light source and the thickness of the specimen are more likely to vary from one image to another. It is possible to make or purchase suitable standards, but fortunately in many cases they are not needed. We do not often need to record images with true colors from microscope slides. The colors often originate with chemical stains, whose colors are known and which are selected for their ability to help us distinguish one structure from another. The actual colors are less important than the fact that they are different in different features. In that case, color standardization is not required.

Thus far, color has been considered as a combination of red, green, and blue. That is the way most cameras work, the way data are stored internally in the computer, and the way that the image is generated on the computer display. A magnifying glass will confirm that all of the colors are made up of little red, green, and blue dots — either phosphors on the cathode ray tube or colored filters on an LCD display. For large stadium-size displays, arrays of red, green, and blue light bulbs may be used. RGB displays cannot generate all of the colors that people can see, but they do well enough for most applications. Technically, the three color phosphors form a triangle in color space that encompasses most of the visible colors, but very saturated colors, especially greens and purples, generally lie outside the gamut. The gamut problem is discussed more fully with regard to printers, where it presents more significant limitations.

People do not perceive color as a combination of red, green, and blue, even though the three kinds of color-sensitive cones in the human eye also respond more or less to red, green, and blue wavelengths. Look around you at the colors on books, clothing, buildings, human skin, and so forth, and you will not find yourself thinking of the amounts of red, green, and blue present. RGB color space is mathematically very convenient but it is not a good space for image processing, either, because small shifts in the proportions of these primary colors result in significant variations in the perceived color of the pixels. Most image processing operations use other systems of coordinates in color space.

COLOR SPACE COORDINATES

Color space is three-dimensional. Fundamentally, this is because human vision has three kinds of light-receptor cones, which are sensitive to different ranges of light wavelengths. Some animals, particularly birds, have evolved eyes with greater spatial resolution and greater color discrimination. Pigeons, for example, possess five kinds of cones and for them, color space would be five dimensional.

RGB color coordinates (Figure 2.14a) provide an attractively simple way to navigate color space. The axes are independent and orthogonal, marking out a cubic volume. The corners of the cube are black, white, red, yellow, green, cyan, blue,

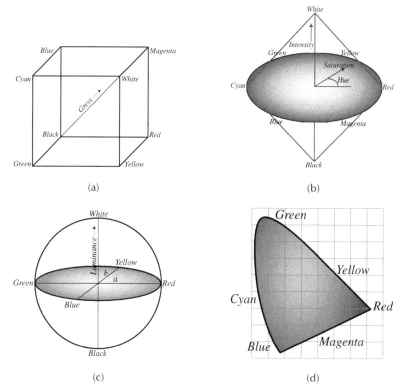

FIGURE 2.14 Several coordinate systems used for color space as discussed in the text: (a) RGB (and CMY) cube, (b) Hue-Saturation-Intensity bi-cone; (c) L-a-b sphere; (d) CIE colors (intensity is measured perpendicular to the plane; fully saturated colors lie around the edge, neutral grey or white is near the center).

and magenta, and the diagonal line from the black to the white corner is the range of grey scale values. RGB color is additive, because the intensities along the red, green, and blue axes add together to produce perceived colors. The same orthogonal axes describe printing technology using cyan, magenta and yellow (CMY) inks. Printing is subtractive, because the white paper provides the background and inks reduce brightness. Combinations of the three inks are plotted along the CMY axes to describe colors within the space.

As noted above, RGB space describes how much of the hardware involved in imaging works, but not how humans perceive color. Other color coordinate systems provide different ways to identify points in this space. They do not map directly onto the RGB cube, but rather involve various distortions of the space that stretch some regions and compress others. Equations that provide the mathematical conversion to and from the various coordinate systems are typically applied by computer programs as needed.

A description of color that seems to correspond well to human perception is HSI (hue-saturation-intensity) There are several variants of these coordinates, which may be plotted as a cylinder, a cone, or a double cone (Figure 2.14b), and may be described as HSB (hue-saturation-brightness), HSL (lightness), or HSV (value). The common characteristic of all these variations is a circular (or approximately circular) hue-saturation map. The center of this circle is a colorless grey scale value. Radial distance from the center point is saturation, a measure of the amount of color present. The difference between pink and red, for example, is saturation. The angle of the direction from the center point is the hue, which is what most people mean by color. This hue-saturation circle is the color wheel children learn about in Kindergarten. The colors progress from red to orange, yellow, green, cyan, blue, magenta, and back to red around the circumference of the circle. The brightness axis is perpendicular to the color wheel.

Most people find describing colors using HSI parameters to be quite understandable, but the spaces are very inconvenient mathematically. The hue angle varies from 0 to 360 degrees starting at red, but a hue of 5 degrees (slightly to the orange of red) and one of 355 degrees (slightly to the magenta of red) average to red, so the angle values must be used modulo 360 in calculations. Furthermore, the maximum saturation varies with brightness. It is not possible to add color to a completely black value without increasing brightness, or to introduce color into a completely white value without reducing at least one of colors and with it the total brightness. But this is the space in which most image processing is best performed.

A similar but mathematically more convenient coordinate system that can also be used is L-a-b, a spherical space with orthogonal coordinates (Figure 2.14c). The vertical axis from the south pole to the north pole of the sphere is luminance, while the a and b axes mark the variation from red to green and from yellow to blue, respectively. Note that the central color wheel in this space is not quite the same as in HSI, but is distorted to place red, yellow, green, and blue at 90 degree intervals around the periphery. A radial line from the grey center would be expected to represent an increase in saturation with no change in hue, but this is only approximately true because of the distortion.

In both HSI and L-a-b coordinates, the blending together of two colors does not produce the colors that lie along a straight line between them. The CIE color coordinates (Figure 2.14d) mentioned previously were designed for that behavior, and, furthermore, there is a modification of CIE coordinates that attempts to represent just-visible-differences between colors as a uniform set of distances throughout the space.

Some of these color coordinates are used primarily for color matching and visual perception, others for control of hardware such as video transmission and color printing, and others for image processing. For our purposes here the distinctions are not important. Whether an image processing operation uses HSI or L-a-b internally matters less than that it does not use RGB. Processing the intensity or luminance value while preserving the color (whether expressed as saturation and hue or a and b values) is generally the best way to enhance detail visibility without altering color perception.

Whichever set of color coordinates is used to describe color space, it must be remembered that different hardware devices (cameras, scanners, displays, printers) do not record or reproduce the same range of colors that human vision can perceive. In some cases, colors beyond the range of human vision can be detected (e.g., silicon detectors are sensitive in the infrared as noted above). In many more situations, the phosphors used in displays and the inks used in printers cannot produce highly saturated colors. The gamut of the device is a measure of the range of colors that can be reproduced.

In order to display and print color images that visually match each other and the original scene that was recorded by the camera, it is important to have data that describe the details of color production. Fortunately, driven by the requirements of other markets such as publishing, advertising, graphic arts, and studio photography, and with the support of companies such as Apple Computer, Hewlett Packard, Epson and others, a solution is now widely implemented. ICC (International Color Consortium) curves describe the color production characteristics of displays and printers, and the files are provided by most manufacturers. Their use is standard in some programs such as Adobe Photoshop and operating systems such as Mac OS X, but not in others such as Powerpoint and Windows 98 (which explains why images that have been adjusted to look great in Photoshop do not preserve the same appearance when placed into Powerpoint presentations).

The difference in gamut between displays and printers can be dealt with in several ways. The two most widely used are to reduce all of the saturation values proportionately so that the image colors fit inside the printer gamut, or to clip those values that fall outside the printer gamut while leaving all other values unchanged. The former method (perceptual intent) has the advantage that differences between colors with differing saturation can be preserved, while the second (relative colorimetric) may be preferred because it does not reduce the overall saturation of the image. The most general advice given is to use relative colorimetric intent in most cases, unless the amount of clipping becomes a significant fraction of the total print area, in which case a shift to perceptual intent will preserve the visibility of detail.

There are three major types of printers commonly available for image reproduction. Dye-sublimation (or dye-diffusion) printers produce the most photograph-like results, with continuous colors but with lower spatial resolution and less saturation than a photographic print. The prints are formed by locally heating a carrier film containing dye, which then diffuses into the coating on special paper. Three dyes, cyan, magenta, and yellow, are used. The diffusion prevents the formation of visible pixel boundaries, and the heating allows proportional control of the amounts of each color, to produce very smooth results, even with only 300 pixels per inch on the print. The drawbacks of this technology are the high cost of both the printers and the consumables (which include special coated paper). If only C, M, and Y dyes are used, there is no true black, just a dark brown. In addition, the process is quite slow. But the resulting prints can be used in place of photographic prints, for example for submitting papers for publication when the publisher does not yet accept electronic submission.

Far more common than dye-sub printers are inkjet printers. These deposit very tiny drops of ink onto the surface of coated papers. Controlling the placement and number of drops of each color produces the visual impression of colors. The best modern printers use droplets of 2 to 4 picoliters with six or more inks, typically black, grey, yellow, and one or two each (light and dark) of cyan, and magenta. With appropriate control software these produce a fairly broad gamut of colors. Ink jet printers are very inexpensive, but the special inks and papers make the per-copy cost rather high, and the process is rather slow. The biggest problems have to do with the inks. Dyes produce the best color fidelity but are not stable for long times, and are particularly sensitive to ultraviolet light and humidity. Pigment based inks are more stable but do not give such good color rendition and have a smaller gamut, and are subject to metamerism (variation in appearance with different lighting or orientation).

Until fairly recently, laser printers were not considered suitable for quality color printing because the deposited toner particles did not produce highly saturated colors or high spatial resolution. That is no longer the case. Laser printers are very fast, use standard papers, and have a low cost per copy (although the original investment in the printer is much greater than an ink jet). The copies are certainly good enough for reports and desk-top publications, but because of the regular half-tone patterns in the colors (cyan, magenta, yellow, and black), the prints do not duplicate well so that it is necessary to print as many original copies as needed.

With all of these printing methods, the white of the paper (or its coating) is one of the important components of the final color. Depositing more ink adds saturation but also darkens the color, so that very bright saturated colors cannot be produced. In most cases it becomes necessary to slightly reduce the total range of intensities in each color channel. The step from pure white paper to the smallest amount of ink or toner can be quite visible, so the brightest white allowed in the image is typically reduced so that it prints with at least a few percent ink coverage. Likewise at the dark end of the brightness range, the halftone dots may become "plugged," meaning that they spread slightly and fill in the spaces between them before the nominal 100% point. Slightly increasing the brightness of the darkest values avoids this problem as well.

There are numerous excellent books that deal in detail with printing technology and preparation of digital images for high quality printout, both on local printers as discussed here, and also for offset press applications. Since Adobe Photoshop is the most common software platform used for handling images in these applications, most of the books deal specifically with that environment. Two fine texts are: Dan Margulis, *Professional Photoshop, the Classic Guide to Color Correction* (John Wiley & Sons, New York) and Michael Kieran, *Photoshop Color Correction* (Peachpit Press, Berkeley). The latter also deals with preparing color images for Web presentation, but that is a difficult task since there is no way to control the settings used by individual browsers, and image appearance can vary greatly.

For most researchers, the need for hard copy prints is limited, and the use of printers with manufacturer-provided ICC curves, and software that knows how to translate the stored RGB values to CMYK (the K stands for black), or other ink combinations suitable for the printer, will produce acceptable results with little special attention. Finally, there are several online service bureaus who will accept uploads of images, print them on photo-quality paper (generally using dye-sub technology) and ship the prints back to you. The quality of these services varies, but for occasional use they may provide a useful supplement to a local laser or ink jet printer.

COLOR CHANNELS

It was pointed out above that images are typically acquired as RGB channels, with 8 or more bits per channel, and are stored that way in most of the various file formats (a few, like PSD, can also store L-a-b or CMYK). The ability of the computer to convert the RGB values to and from other color coordinates makes it possible to arbitrarily change the interpretation of the image from one color space to another whenever it is convenient. In particular, it is often advantageous to merge separate images as color channels, to split the color channels into separate images, or to insert a grey scale image into a color image in any chosen color channel.

Merging separate images of the same area acquired through different filters is often an effective way to produce a color image as shown in Figure 2.15. Typically, the channels are assigned to the RGB channels even though they may not actually represent those colors — or any colors. Multiple images from the confocal microscope using different laser excitation of fluorescent dyes may include wavelengths in the UV or infrared. Images from the SEM can include secondary and backscattered electrons, electron signals collected from different directions in the chamber, and, of course, X-ray signals corresponding to various elements in the periodic table.

Assigning these three at a time produces a color image in which the proportion of the various signals at disparate points in the image can be judged by the color and brightness of the combined colors. When there are more than three signals available, it is not possible to combine them in a single image. For example, if red, green, and blue have been assigned to three signals, you can not assign a fourth signal to yellow, because the combination of green and red will produce yellow wherever both signals are present in equal proportions. To handle more than three signals at once we would need a visual system like the pigeon with more than three types of color receptors. Selecting the three signals that best represent the structure, and assigning them to the colors that portray this most effectively for communicating

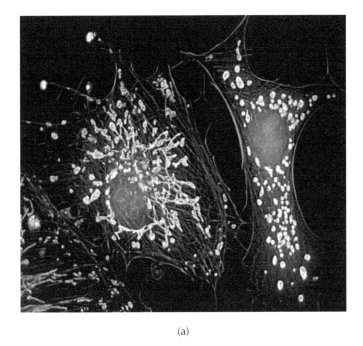

(a)

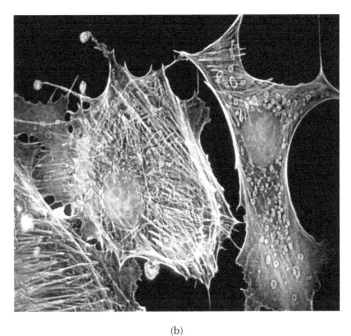

(b)

FIGURE 2.15 Merging of separately acquired color channel images from a confocal light microscope: (a) red channel; (b) green channel; (c) blue channel; (d) merged result (shown in Color Figure 2.15; see color insert following page 150).

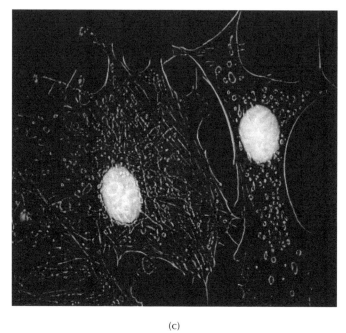

(c)

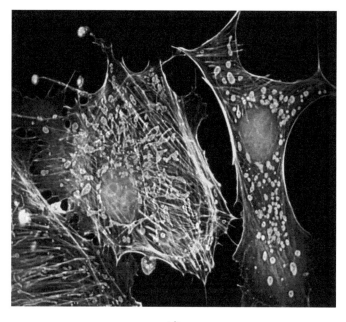

(d)

FIGURE 2.15 (continued)

the results to someone else who has not studied the original images, is something of an art. There is no best way to do it, although there are often a lot of not so good ways.

Channel merging is also a standard way to present stereo pair images. Colored glasses with the red lens on the left eye and either green or blue on the right eye are often used to view composite images which are prepared by placing the corresponding eye views into the color channels.

Separating the channels can also be an aid to visual perception of the information in an image. The presence of color is important in human vision but not as important as variations in brightness. The reason that the various compression methods described above reduce the spatial and tonal resolution of the color information more than the brightness is that human vision does not generally notice that reduction. Broadcast television and modern digital video recording use this same trick, assigning twice the bandwidth for the brightness values as for the color information. Blurring the color information in an image so that the colors bleed across boundaries does not cause any discomfort on the part of the viewer (which may be why children do not always color inside the lines).

Seeing the important changes in brightness in an image, which often defines and locates important features, may be easier in some cases if the distracting variations in color, or the presence of a single dominant color, are removed. In the color original of meat shown in Figure 2.2 the predominant color is red. Examining the red channel as a monochrome (grey scale) image shows little contrast because there is red everywhere. On the other hand, the green channel (or a mixture of green and blue, as used in the example) shows good contrast. In general a complementary hue, opposite on the color wheel, will reveal information hidden by the presence of a dominant color (just as a photographer uses a yellow filter to enhance the visibility of clouds in a photograph of blue sky). Of course, in this example an equivalent result could have been obtained by recording the image with a monochrome camera through a green filter, rather than acquiring the color image and then extracting the green channel.

In other cases it is the hue or saturation channels that are most interesting. When colored stains or dyes are introduced into biological material (Figure 2.16) in order to color particular structures or localize chemical activity, the colors are selected so they will be different. That difference is a difference in hue, and examining just the hue channel will show it clearly. Likewise, the saturation channel intensity corresponds to the amount of the stain in each location. The intensity channel records the variation in density of the specimen.

There are many ways to extract a monochrome (grey scale) image from a color image, usually with the goal of providing enhanced contrast for the structures present. The simplest method is to simply average the red, green, and blue channels. This does not correspond to the brightness that human vision perceives in a color scene, because the human eye is primarily sensitive to green wavelengths, less to red, and much less to blue. Blending the channels in proportions of about 65% green, 25% red, 10% blue will give approximately that result.

But it is also possible to mix the channels arbitrarily to increase the contrast between particular colors. An optimum contrast grey image can be constructed from any color image by fitting a regression line through a 3D plot of the pixel color

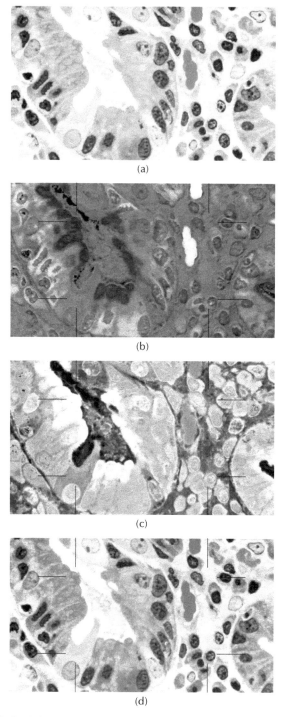

(a)

(b)

(c)

(d)

FIGURE 2.16 Stained tissue imaged in the light microscope, and the HSI color channels: (a) color original (see color insert following page 150); (b) hue; (c) saturation; (d) intensity.

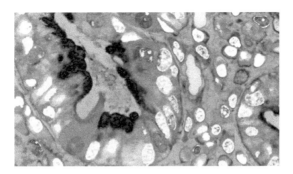

FIGURE 2.17 Mixing the RGB color channels in the color image from Figure 2.16(a) to produce the optimum grey scale contrast.

values and mapping the grey scale values along that line (Figure 2.17). In principle this can be done in any color coordinates, but it is easiest to carry out and to visualize in RGB.

OPTIMUM IMAGE CONTRAST

Whether images are color or monochrome (many imaging techniques such as SEM, TEM, AFM, etc. produce only grey scale images), it is desirable to have the best possible image contrast. Ideally, this can be done by controlling the illumination, camera gain and contrast, digitizer settings, etc., so that the captured image covers the full available brightness range and the discrimination between structures is optimum. In practice, images are often obtained that can be improved by subsequent processing.

The first tool to understand and use for examining images is the histogram. This is usually presented as a plot of the number of pixels having each possible brightness value. For an 8 bit greyscale image, with only 256 possible brightness values, this plot is compact and easy to view. For an image with greater bit depth, the plot would be extremely wide and there would be very few counts for most of the brightness values, so it is common to bin them together into a smaller number of levels (usually the same 256 as for 8 bit images). For a color image, individual histograms may be presented for the various channels (RGB or HSI) as shown in Figure 2.18, or other presentations showing the distribution of combinations of values may be used. It generally takes some experience to use color histograms effectively, but in most cases the desired changes in contrast can be effected by working just with the intensity channel, leaving the color information unchanged.

A histogram that shows a significant number of pixels at the extreme white or black ends of the plot (as shown in Figure 2.19) indicates that the image has been captured with too great a contrast or gain value, and that data have been lost due to clipping (values that exceed the limits of the digitizer are set to those limits). This situation must be avoided because data lost to clipping cannot be recovered, and it

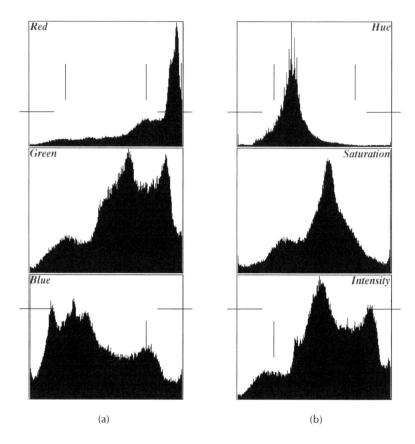

(a)   (b)

FIGURE 2.18 Histograms for the RGB and HSI channels of the stained tissue image in Figure 2.16. In each, the horizontal axis goes 0 (dark) to 255 (maximum), and the vertical axis is the number of pixels.

is likely that the size and other details of the remaining features have been altered. This problem frequently arises when particulates are dispersed on a bright or dark substrate. Adjusting contrast so that the substrate brightness is clipped reduces the size of the particles.

A histogram that does not cover the full range of intensity values from light to dark indicates that the image has low contrast (Figure 2.20). Simply stretching the values out to cover the full range will leave gaps throughout the histogram, but because the human eye cannot detect small brightness differences this is not a problem for visual examination. There are just as many missing values in the stretched histogram as in the original, but in the original they were grouped together at the high and low brightness end of the scale where their absence was noticeable. Human vision can detect a brightness change of several percent under good viewing conditions, which translates to roughly 30 brightness levels in a scene. Most modern computer monitors can display 256 brightness levels, so missing a few scattered throughout the range does not degrade the image appearance.

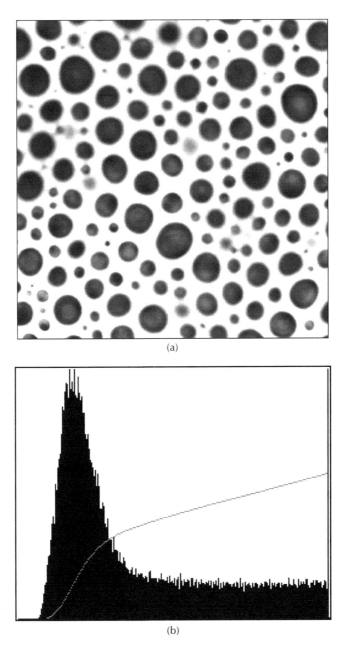

(a)

(b)

FIGURE 2.19 Bubbles in a whipped food product: (a) image (courtesy of Allen Foegeding, North Carolina State University, Department of Food Science) in which much of the white area around the bubbles is pure white (value = 255); (b) histogram, showing a large pixel count at 255 (white) where the single line in the histogram rises off scale. The cumulative histogram (grey line) is the integral of the histogram. Since 100% is full scale, the cumulative plot indicates that about 40% of the image area is occupied by exactly white pixels.

Human vision is not linear, but instead detects a proportional change in brightness. Equal increments in brightness are viewed differently at the bright and dark ends of the brightness range. Adjusting the overall gamma value of the image allows detail to be seen equally well in bright and dark regions. One effective way to control gamma settings is by setting the brightness level (on the 0 to 255 scale) for the midpoint of the pixel brightness range (the median value, with 50% darker and 50% lighter). This produces a smooth curve relating the displayed brightness value to the original value, which can either stretch out dark values to increase detail visibility in dark regions by compressing bright values so detail there becomes less distinguishable, or vice versa.

For the experienced film photographer, setting the midpoint of the brightness range to a desired value is equivalent to the zone system for controlling the contrast of prints. Typical film or digital camera images can record a much wider latitude of brightness values than can be printed as hardcopy, and selecting a target zone for the image directly affects the visibility of detail in the bright and dark areas of the resulting image as shown in Figure 2.21.

Photographers with darkroom experience also know that detail is often visible in the negative that is hard to see in the print, and vice versa. Because of the nonlinearity of human vision, simply inverting the image contrast often assists in visual perception of detail in images.

Of course, there is no reason that the curve relating displayed brightness to original brightness (usually called the "transfer function") must be a smooth constant-gamma curve. Arbitrary relationships can be used to increase brightness in any particular segment of the total range, and these may even include reversal of contrast in one part of the brightness range, equivalent to the darkroom technique of solarization. Figure 2.22 illustrates a few possibilities.

A specific curve can be used as a transfer function for a given image that transforms the brightness values so that equal areas of the image are displayed with each brightness value. This is histogram equalization, and is most readily visualized by replotting the conventional histogram as a cumulative plot of the number of pixels with brightness values equal to or less than each brightness values (mathematically, the integral of the usual plot). As shown in Figure 2.23, this curve rises quickly where there is a peak and slowly where there is not. If the cumulative histogram curve is used as the transfer function, it spreads out subtle gradients and makes them more visible. It also transforms the cumulative histogram so that it becomes a straight line.

Transforms like these that alter the brightness values of pixels in the image are global, because each pixel is affected the same way. That is, two pixels that started out with the same value will end up with the same value (although it will probably be a different one from the original) regardless of where they are located or what their surroundings may be. The goal is to increase the ability of the viewer to perceive the contrast in the image. To achieve this, any consistent relationship between brightness and a physical property of the specimen is sacrificed. Density, dosage measurements, activity, or other properties that might be calibrated against brightness cannot be measured after these brightness transformations.

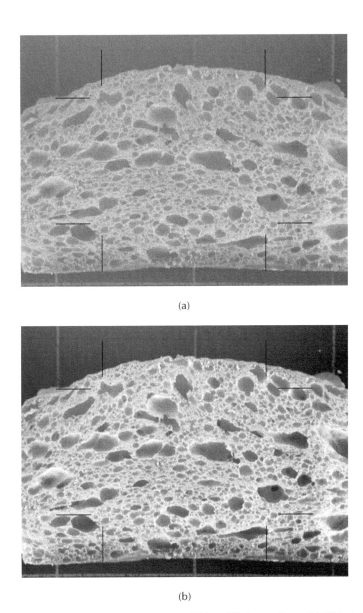

(a)

(b)

FIGURE 2.20 Image of bread slice (courtesy of Diana Kittleson, General Mills): (a) original low contrast image; (b) contrast stretched linearly to maximum; (c) original histogram showing black and white limits set to end points; (d) histogram after stretching to full range.

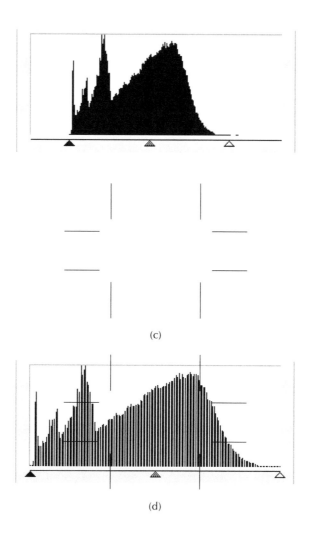

(c)

(d)

FIGURE 2.20 (continued)

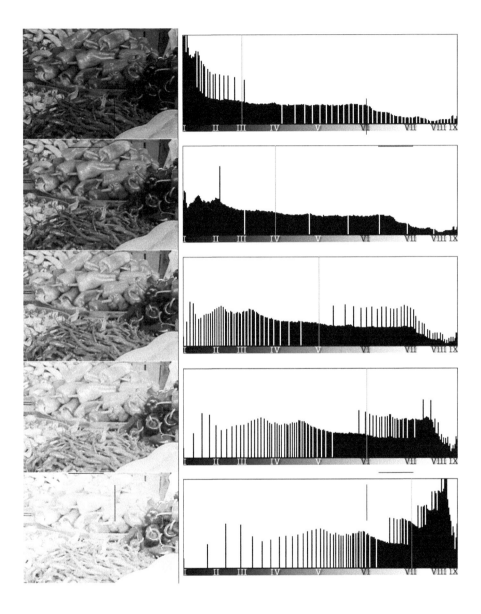

FIGURE 2.21 Adjusting the appearance of an image by setting the target zone (from III to VII) for the midpoint of the histogram. Note the appearance of gaps and pileup at either end of the histogram (fragment of the market image in Color Figure 2.3; the histogram represents the entire image; see color insert following page 150).

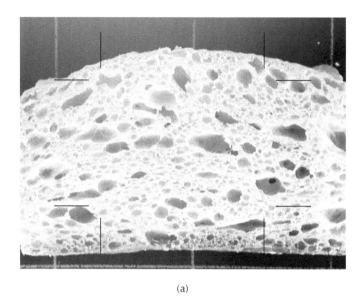

(a)

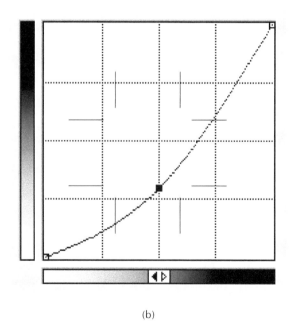

(b)

FIGURE 2.22 Manipulation of the transfer function for the bread slice image from Figure 2.20: (a, b) positive gamma reveals detail inside the dark shadow areas; (c, d) negative gamma increases the contrast in the bright areas; (e, f) inverting the contrast produces a negative image; (g, h) solarization shows both positive and negative contrast. The transfer function or contrast curve used for each image is shown; the horizontal axis shows the stored brightness values and the vertical axis shows the corresponding displayed brightness.

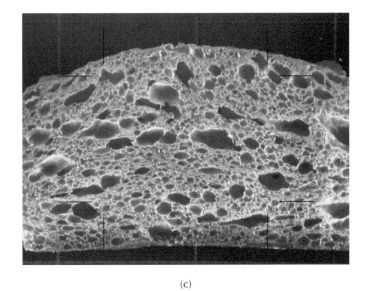

(c)

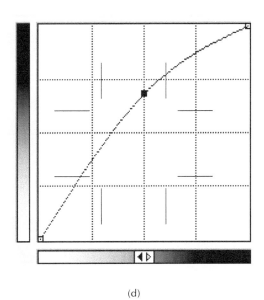

(d)

FIGURE 2.22 (continued)

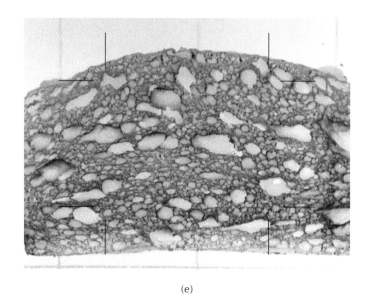

(e)

(f)

FIGURE 2.22 (continued)

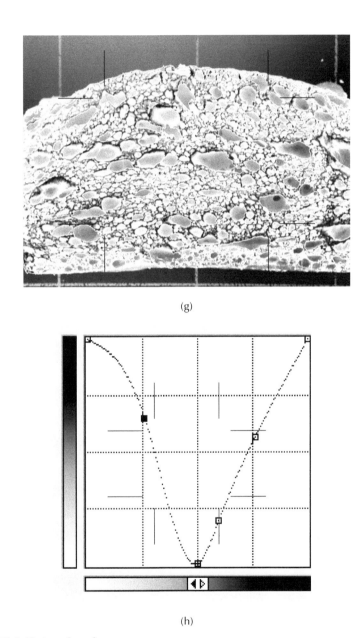

(g)

(h)

FIGURE 2.22 (continued)

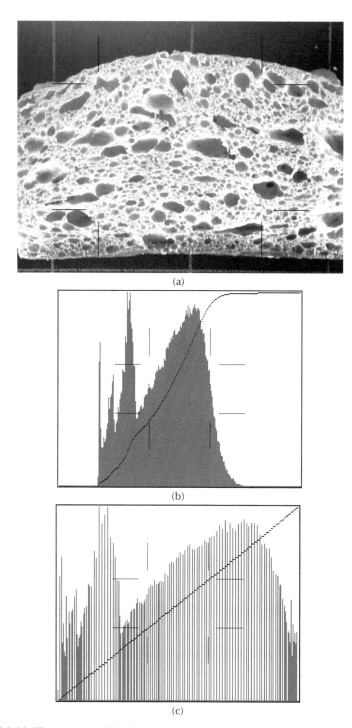

(a)

(b)

(c)

FIGURE 2.23 Histogram equalization: (a) result when applied to original image in Figure 2.20; (b) original histogram showing cumulative histogram used as transfer function; (c) resulting histogram, showing linear cumulative histogram.

REMOVING NOISE

Noise means many different things in different circumstances. Generally, it refers to some part of the image signal that does not represent the actual subject but has been introduced by the imaging system. In film photography, the grains of halide particles or the dye molecules may be described as "noise superimposed on the real image." So, too, may dust or scratches on the film. There are rough equivalents to these with digital photography, and some other problems as well.

A digital picture of a perfectly uniform and uniformly illuminated test card does not consist of identical pixel values. The histogram of such an image typically shows a peak with a generally Gaussian shape. The width of the peak is a measure of the random variation or speckle in the pixels, which although much coarser than the grain in film, has the same underlying statistical characteristics. All of the processes of generating charge in the detectors, transferring that charge out of the chip, amplifying the analog voltage that results, and digitizing that voltage, are statistical in nature. Some, such as the efficiency of each transistor, also vary because of the finite tolerances of manufacture. For scanning microscopes, there are also both statistical variations such as the production of X-rays or secondary electrons and stability concerns such as the electron gun emission or creep in piezo drivers.

The statistical variations are sensitive to the magnitude of the signal. In the SEM, increasing the beam current or slowing the scan rate down so that more electrons are collected at each point, reduces the amount of the speckle as a percentage of the value, and so reduces the width of the peak in the histogram as shown in Figure 2.24. Instrumentation variations do not reduce in magnitude as the signal increases. Some, such as transistor variations on the camera chip, do not vary with time, and are described as "fixed pattern noise." The two types of noise are multiplicative noise (proportional to signal) and additive noise (independent of signal), respectively.

Consider a test image consisting of greyscale steps (a similar image will be used in Chapter 5 to calibrate measurements based on brightness). If an image contains only additive noise, and if the detector is linear (output proportional to intensity) the width of each peak corresponding to one of the grey scale steps would be the same. If the image contains only multiplicative noise, the peak widths would be proportional to the intensity value. Figure 2.24 illustrates both situations. This changes if the final output is logarithmic, in which case multiplicative noise produces equal width peaks. In practice, most devices are subject to both types of noise and the peak widths will vary with brightness but not in strict proportion. One of the important characteristics of random noise is that the amount of noise in an image varies from bright to dark areas, and some noise reduction schemes take this into account.

Although the absolute amount of random noise may increase with signal, it usually drops as a percentage of the signal. Collecting more signal by averaging over time adds to the signal while the random variations may be either positive or negative, and tend to cancel out. For purely multiplicative noise, the signal-to-noise ratio increases as the square root of the averaging time. Of course, in many real situations the strategy of temporal averaging (collecting more signal) to reduce noise is not practical or even possible.

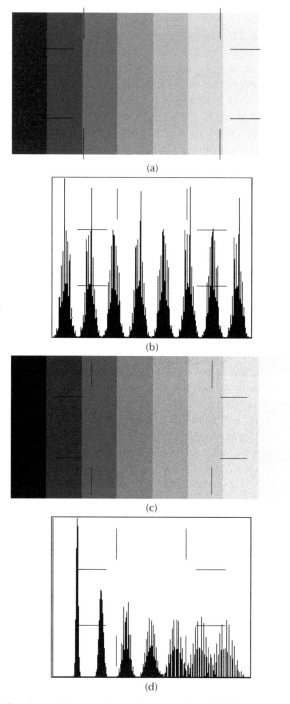

(a)

(b)

(c)

(d)

FIGURE 2.24 Random noise superimposed on a series of brightness steps, and the image histograms: (a, b) additive noise; (c, d) multiplicative noise.

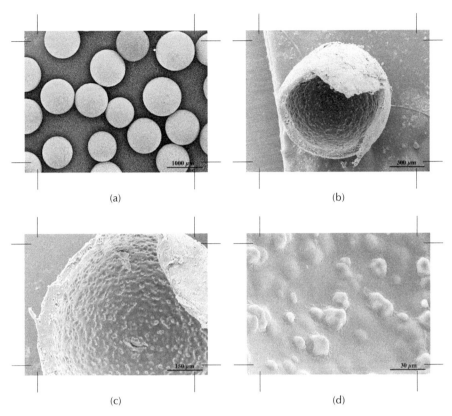

FIGURE 2.25 SEM images at increasing magnification of chewing gum, showing the small flavor crystals embedded on the inside surface of the hollow chicle spheres. (Courtesy of Pia Wahlberg, Danish Technological Institute)

In most images, the size of features is much larger than the size of individual pixels. That means that most of the pixels represent the same structure as their neighbors, and should have the same value. So if temporal averaging to reduce noise is impractical, spatial averaging should also offer a way to reduce noise. Figure 2.25 will be used to illustrate this procedure, and several others described below. A 7 pixel wide circular neighborhood (actually an octagon, the closest approximation to a circle possible on a square pixel grid) is moved across the image. At every position, the values of the pixels in the neighborhood are averaged together and the result is put into the pixel at the center of the circle (Figures 2.26b and 2.27b).

This is an example of a neighborhood operation. Unlike global operations, such as the histogram stretching techniques described above, neighborhood operations are a second major category of image processing operations. With this category of procedures, two pixels that initially have the same value may end up with different

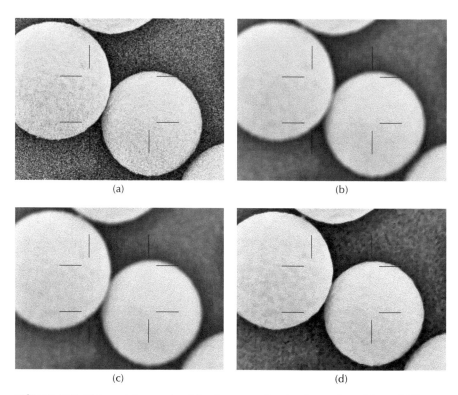

(a)   (b)

(c)   (d)

FIGURE 2.26 Enlarged fragments of the low magnification image from Figure 2.25 comparing the results of noise reduction methods: (a) original image; (b) averaging pixel values in a 7 pixel wide circle; (c) Gaussian smooth with standard deviation of 2 pixel radius; (d) median filter with a 7 pixel wide neighborhood.

values if they have different neighborhoods. Neighborhood operations always use the original pixel values from the image to calculate new values, not the newly calculated values for other pixels in the neighborhood that have already been processed.

Notice in the image that the noise is reduced as expected, but the edges of features have become blurred. The assumption that a pixel should be like its neighbors is not met for pixels near feature edges. The averaging procedure blurs edges, and can also shift their position or alter the shape which makes subsequent delineation and measurement of features difficult.

Instead of simple averaging in which all pixels in the neighborhood are added equally, Gaussian smoothing uses weights for the pixels that vary with distance from the center pixel. Plotting the weight values used in the kernel shows that they describe a Gaussian or bell-shaped curve (Figure 2.28). The standard deviation of the Gaussian controls the amount of smoothing and noise reduction. Ideally the weight values should be real numbers and the neighborhood diameter should be about six times

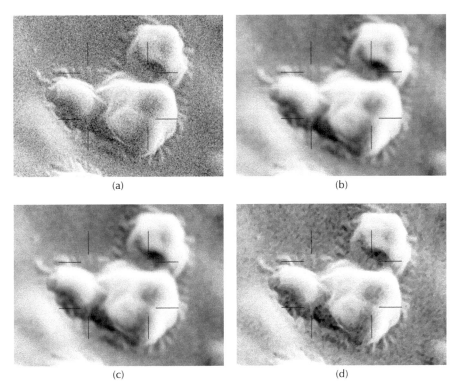

FIGURE 2.27 Enlarged fragments of the high magnification image from Figure 2.25 comparing the results of noise reduction methods: (a) original image; (b) averaging pixel values in a 7 pixel wide circle; (c) Gaussian smooth with standard deviation of 2 pixel radius; (d) median filter with a 7 pixel wide neighborhood.

the standard deviation, but approximations with integer values and size truncation are often used. Practically all image processing programs include Gaussian smoothing. It can be implemented in a particularly efficient way by separately smoothing in the horizontal and vertical directions. Gaussian smoothing (Figures 2.26c and 2.27c) produces the least amount of edge blurring for a given amount of noise reduction, but does not eliminate the problem.

This type of smoothing with a neighborhood kernel of weights is a low-pass filter, because it tends to remove high frequencies (rapid local variations of brightness) in the image while not affecting low frequencies (gradual overall variations in brightness). This raises the important point that much image processing is actually performed in frequency space with Fourier transforms, rather than in the pixel space of the original values.

0.0111	0.0388	0.0821	0.1054	0.0821	0.0388	0.0111
0.0388	0.1353	0.2865	0.3679	0.2865	0.1353	0.0388
0.0821	0.2865	0.6065	0.7788	0.6065	0.2865	0.0821
0.1054	0.3679	0.7788	1.0000	0.7788	0.3679	0.1054
0.0821	0.2865	0.6065	0.7788	0.6065	0.2865	0.0821
0.0388	0.1353	0.2865	0.3679	0.2865	0.1353	0.0388
0.0111	0.0388	0.0821	0.1054	0.0821	0.0388	0.0111

(a)

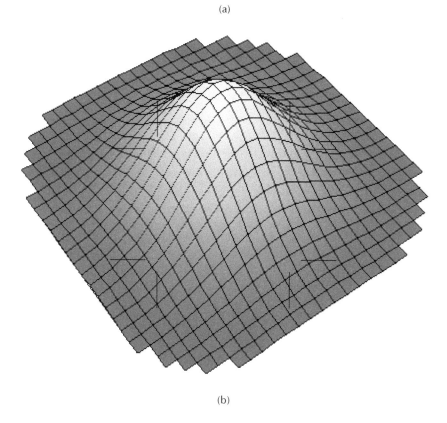

(b)

FIGURE 2.28 A Gaussian smoothing kernel consists of real number weights (or integer approximations) that model the shape of a bell curve: (a) central portion of a Gaussian smoothing kernel, weight values entered into a program dialog; (b) plot of the values for the full kernel, which is 25 × 25 pixels in size.

Fourier transforms will be encountered several times in subsequent chapters on enhancement and measurement. The basis for all of these procedures is Fourier's theorem, which states that any signal (such as brightness as a function of position) can be constructed by adding together a series of sinusoids with different frequencies, by adjusting the amplitude and phase of each one. Calculating those amplitudes and phases with a Fast Fourier Transform (FFT) generates a display, the power spectrum, that shows the amplitude of each sinusoid as a function of frequency and orientation. Most image processing texts (see J. C. Russ, *The Image Processing Handbook*, 4th edition, CRC Press, Boca Raton, FL) include extensive details about the mathematics and programming procedures for this calculation. For our purposes here the math is less important than observing and becoming familiar with the results.

For a simple image consisting of just a few obvious sinusoids, the power spectrum consists of just the corresponding number of points (each spike is shown twice because of the rotational symmetry of the plot). For each one, the radius from the center is proportional to frequency and the angle from the center identifies the orientation. In many cases it will be easier to measure spacings and orientations of structures from the FFT power spectrum, and it is also easier to remove one or another component of an image by filtering or masking the FFT. In the example shown in Figure 2.29, using a mask or filter to set the amplitude to zero for one of the frequencies and then performing an inverse FFT removes the corresponding set of lines without affecting anything else.

A low pass filter like the Gaussian keeps the amplitude of low frequency sinusoids unchanged but reduces and finally erases the amplitude of high frequencies (large radius in the power spectrum). For large standard deviation Gaussian filters, it is more efficient to actually execute the operation by performing the FFT, filtering the data there, and performing an inverse FFT as shown in Figure 2.30, even though it may be easier for those not familiar with this technique to understand the operation based on the kernel of neighborhood weights. Mathematically, it can be shown that these two ways of carrying out the procedure are identical.

There is another class of neighborhood filters that do not have equivalents in frequency space. These are ranking filters that operate by listing the values of the pixels in the neighborhood in brightness order. From this list it is possible to replace the central pixel with the darkest, lightest or median value, for example. The median filter, which uses the value from the middle of the list, is also an effective noise reducer. Unlike the averaging or Gaussian smoothing filters, the median filter does not blur or shift edges, and is thus generally preferred for purposes of reducing noise. Figures 2.26d and 2.27d include a comparison of median filtering with neighborhood smoothing.

Increasing the size of the neighborhood used in the median is used to define the size of details — including spaces between features — that are defined as noise, because anything smaller than the radius of the neighborhood cannot contribute the median value and is eliminated. Figure 2.31 illustrates the effect of neighborhood size on the median filter. Since many cameras provide images with more pixels than the actual resolution of the device provides, it is often important to eliminate noise

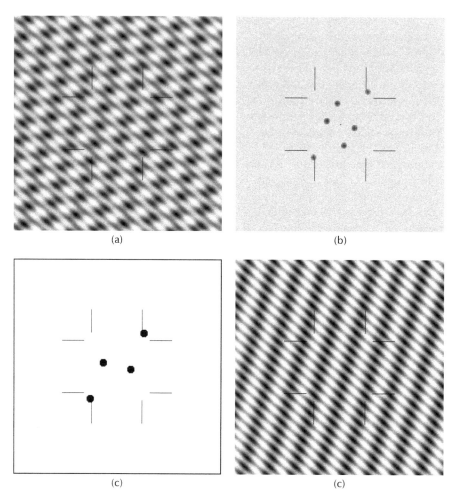

FIGURE 2.29 Illustration of Fourier transform, and the relationship between frequency patterns in the pixel domain and spikes in the power spectrum: (a) image produced by superposition of three sets of lines; (b) FFT power spectrum of the image in (a), showing three spikes (each one plotted twice with rotational symmetry); (c) mask used to select just two of the frequencies; (d) inverse FFT using the mask in (c), showing just two sets of lines.

that covers several pixels. Retention of fine detail requires a small neighborhood size, but because it does not shift edges, a small median can be repeated several times to reduce noise. If the noise is not isotropic, such as scan line noise from video cameras or AFMs, or scratches on film, the use of a neighborhood that is not circular but instead is shaped to operate along a line perpendicular to the noise, the median can also be an effective noise removal tool (see Figure 2.32).

One problem with the median filter is that while it does not shift or blur edges, it does tend to round corners and to erase fine lines (which, if they are narrower

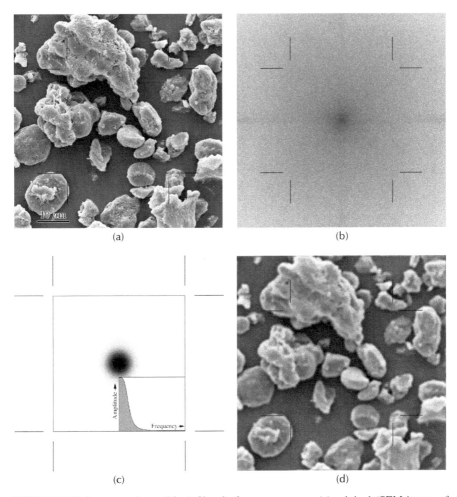

FIGURE 2.30 Low pass (smoothing) filter in frequency space: (a) original (SEM image of spray dried soy protein isolate particles); (b) FFT power spectrum; (c) filter that keeps low frequencies and attenuates high frequencies; (d) inverse FFT produces smoothed result.

than the radius of the neighborhood, are considered to be noise). This can be corrected by using the hybrid median. Instead of a single ranking operation on all of the pixels in the neighborhood, the hybrid median performs multiple rankings on subsets of the neighborhood. For the case of the 3×3 neighborhood, the ranking is performed first on the 5 pixels that form a + pattern, then the five that form an x, and finally on the original central pixel and the median results from the first two rankings. The final result is then saved as the new pixel value. This method can be extended to larger neighborhoods and more subsets in additional orientations. As shown in the example in Figure 2.33c, fine lines and sharp corners are preserved.

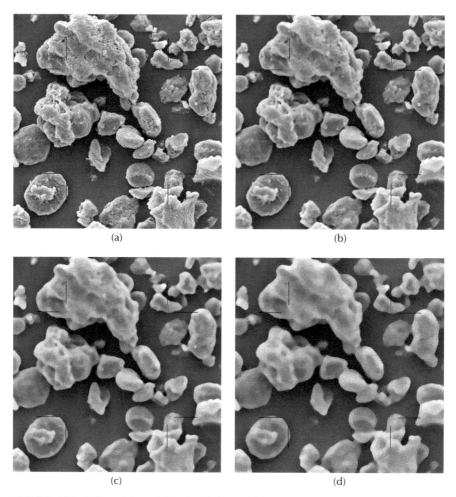

(a)

(b)

(c)

(d)

FIGURE 2.31 Effect of the neighborhood size used for the median filter: (a) original image; (b) 5 pixel diameter; (c) 9 pixel diameter; (d) 13 pixel diameter.

Another approach to modifying the neighborhood is the conditional median (Figure 2.33d). In addition to the radius of a circular neighborhood, the user specifies a threshold value. Pixels whose difference from the central pixel exceed the threshold are not included in the ranking operation. This technique also works to preserve fine lines and irregular shapes.

All of these descriptions of a median filter depend on being able to rank the pixels in the neighborhood in order of brightness. Their application to a grey scale image is straightforward, but what about color images? In many cases with digital cameras, the noise content of each channel is different. The blue channel in particular generally has a higher noise level than the others because silicon detectors are

 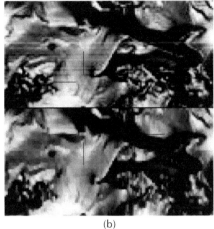

(a)   (b)

FIGURE 2.32 Removal of scan line noise in AFM images: (a) original image (surface of chocolate); (b) application of a median filter using a 5 pixel vertical neighborhood to remove the scan line noise (top) leaving other details intact (bottom).

relatively insensitive to short wavelengths and more amplification is required. The blue channel also typically has less resolution than the green channel (half as many blue filters are used in the Bayer pattern). Separate filtering of the RGB channels using different amounts of noise reduction (e.g., different neighborhood sizes) may be used in these cases.

The problem with independent channel filtering is that it can alter the proportions of the different color signals, resulting in the introduction of different colors that are visually distracting. It is usually better to filter only the intensity channel, leaving the color information unchanged. Another approach uses the full color information in a median filter. One approach is to use a brightness value for each pixel, usually just the sum of the red, green and blue values, to select the neighbor whose color values replace those of the central pixel. This does not work as well as performing a true color median, although it is computationally simpler.

Finding the median pixel using the color values requires plotting each pixel in the neighborhood as a point in color space, using one of the previously described systems of coordinates. It is then possible to calculate the sum of distances from each point to all of the others. The median pixel is the one whose point has the smallest sum of distances, in other words is closest to all of the other points. It is also possible and equivalent to define this point in terms of the angles of vectors to the other points. In either case, the color values from the median point are then reassigned to the central pixel in the neighborhood. The color median can be used in conjunction with any of the neighborhood modification schemes (hybrid median, conditional median, etc.).

Random speckle noise is not the only type of noise defect present in images. One other, scratches, has already been mentioned. Most affordable digital cameras suffer from a defect in which a few detectors are inactive (dead) or their output is

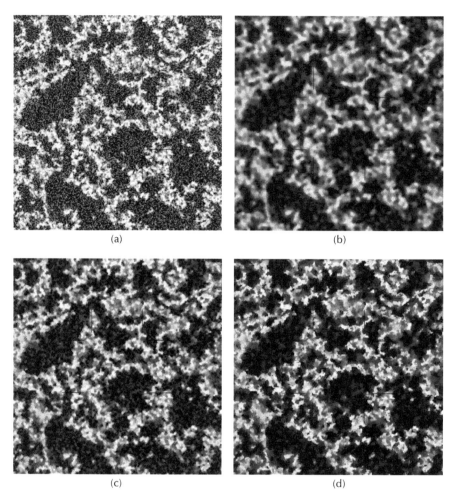

(a)   (b)

(c)   (d)

FIGURE 2.33 Comparison of standard and hybrid median: (a) original (noisy CSLM image of acid casein gel, courtesy of M. Faergemand, Department of Dairy and Food Science, Royal Veterinary and Agricultural University, Denmark); (b) conventional median; (c) hybrid median; (d) conditional median.

always maximum (locked). This produces white or black values for those pixels, or at least minimum or maximum values in one color channel. A similar problem can arise in some types of microscopy such as AFM or interference microscopes, where no signal is obtained at some points and the pixels are set to black. Dust on film can produce a similar result. Generally, this type of defect is shot noise.

A smoothing filter based on the average or Gaussian is very ineffective with shot noise, because the errant extreme value is simply spread out into the surrounding pixels. Median filters, on the other hand, eliminate it easily and completely, replacing the defective pixel with the most plausible value taken from the surrounding neighborhood, as shown in Figure 2.34.

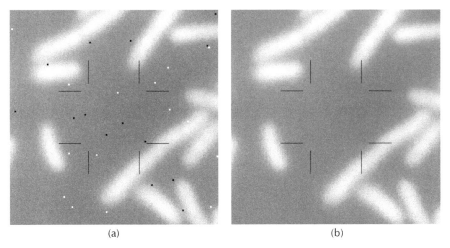

(a)   (b)

FIGURE 2.34 Removal of shot noise: (a) original image of bacteria corrupted with random black and white pixels; (b) application of a hybrid median filter.

Periodic noise in images shows up as a superimposed pattern, usually of lines. It can be produced by electronic interference or vibration, and is also present in printed images such as the ones in this book because they are printed with a halftone pattern that reproduces the image as a set of regularly spaced dots of varying size. Television viewers will recognize moiré patterns that arise when someone wears clothing whose pattern is similar in size to the spacing of video scan lines as a type of periodic noise that introduces strange and shifting colors into the scene, and the same phenomenon can happen with digital cameras and scanners.

The removal of periodic noise is practically always accomplished by using frequency space. Since the noise consists of just a few, usually relatively high frequencies, the FFT represents the noise pattern as just a few points in the power spectrum with large amplitudes. These spikes can be found either manually or automatically, using some of the techniques described in the next chapter. However they are located, reduction of the amplitude to zero for those frequencies will eliminate the pattern without removing any other information from the image, as shown in Figure 2.35.

Because of the way that single chip color cameras use filters to sample the colors in the image, and the way that offset printing uses different halftone grids set at different angles to reproduce color images, the periodic noise in color images is typically very different in each color channel. This requires processing each channel separately to remove the noise, and then recombining the results. It is important to select the right color channels for this purpose. For instance, RGB channels correspond to how most digital cameras record color, while CMYK channels correspond to how color images are printed.

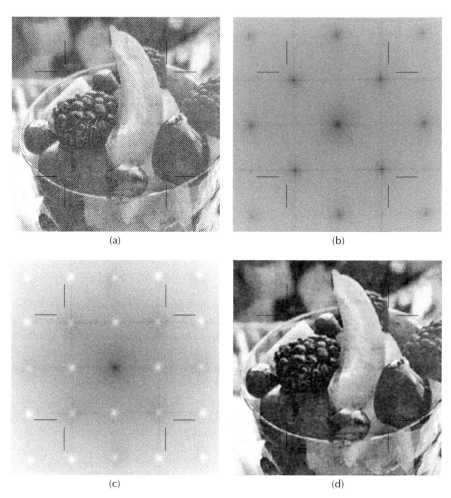

(a)   (b)

(c)   (d)

FIGURE 2.35 Removal of periodic noise: (a) image from a newspaper showing halftone printing pattern; (b) Fourier transform power spectrum with spikes corresponding to the high frequency pattern; (c) filtering of the Fourier transform to remove the spikes; (d) result of applying an inverse Fourier transform.

NONUNIFORM ILLUMINATION

One assumption that underlies nearly all steps in image processing and measurement, as well as strategies for sampling material, is that the same feature will have the same appearance wherever it happens to be positioned in an image. Nonuniform illumination violates this assumption, and may arise for a number of different causes. Some of them can be corrected in hardware if detected before the images are acquired, but some cannot and we are often faced with the need to deal with previously recorded images that have existing problems.

Variations that are not visually apparent (because the human eye compensates automatically for gradual changes in brightness) may be detected only when the image is captured in the computer. Balancing lighting across the entire recorded scene is difficult. Careful position of lights on a copy stand, use of ring lighting for macro photography, or adjustment of the condenser lens in a microscope, are procedures that help to achieve uniform lighting of the sample. Capturing an image of a uniform grey card or blank slide and measuring the brightness variation is an important tool for such adjustments.

Some other problems are not normally correctable. Optics can cause vignetting (darkening of the periphery of the image) because of light absorption in the glass. Cameras may have fixed pattern noise that causes local brightness variations. Correcting variations in the brightness of illumination may leave variations in the angle or color of the illumination. And, of course, the sample itself may have local variations in density, thickness, surface flatness, and so forth, which can cause changes in brightness.

Many of the variations other than those which are a function of the sample itself can be corrected by capturing an image that shows just the variation. Removing the sample and recording an image of just the background, a grey card, or a blank slide or specimen stub with the same illumination provides a measure of the variation. This background image can then be subtracted from or divided into the image of the sample to level the brightness. The example in Figure 2.36 shows particles of cornstarch imaged in the light microscope with imperfect centering of the light source. Measuring the particle size distribution depends upon leveling the contrast so that particles can be thresholded everywhere in the image. Capturing a background image with the same illumination conditions and subtracting it from the original makes this possible.

The choice of subtraction or division for the background depends on whether the imaging device is linear or logarithmic. Scanners are inherently linear, so that the measured pixel value is directly proportional to the light intensity. The output from most scanning microscopes is also linear, unless nonlinear gamma adjustments are made in the amplified signal. The detectors used in digital cameras are linear, but in many cases the output is converted to logarithmic to mimic the behavior of film. Photographic film responds logarithmically to light intensity, with equal increments of density corresponding to equal ratios of brightness. For linear recordings, the background is divided into the image, while for logarithmic images it is subtracted (since division of numbers corresponds to the subtraction of their logarithms). In practice, the best advice when the response of the detector is unknown, is to try both methods and use the one that produces the best result. In the examples that follow, some backgrounds are subtracted and some are divided to produce a level result.

In situations where a satisfactory background image cannot be (or was not) stored along with the image of the specimen, there are several ways to construct one. In some cases one color channel may contain little detail but may still serve as a measure of the variation in illumination. Another technique that is sometimes used is to apply an extreme low pass filter (e.g., a Gaussian smooth with a large standard deviation) to the image to remove the features, leaving just the background variation. This method is based on the assumptions that the features are small compared to

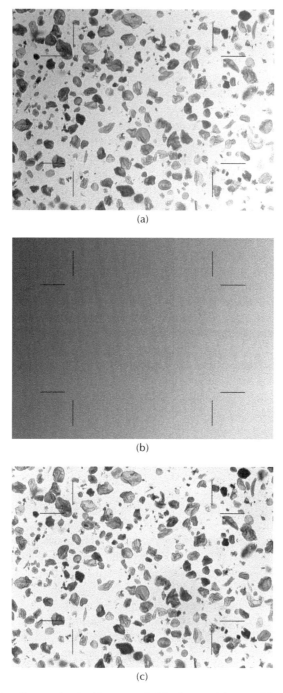

(a)

(b)

(c)

FIGURE 2.36 Leveling contrast with a recorded background image: (a) original image of cornstarch particles with nonuniform illumination; (b) image of blank slide captured with same illumination; (c) subtraction of background from original.

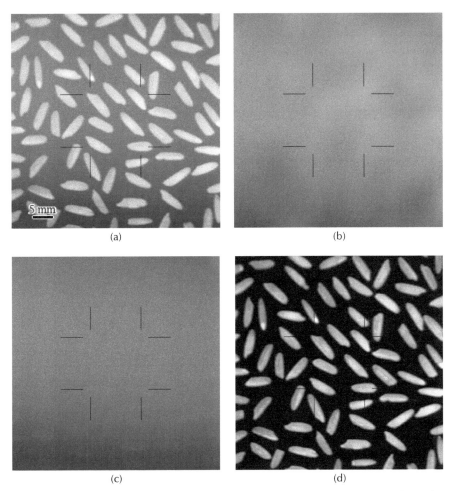

(a)   (b)

(c)   (d)

FIGURE 2.37 Removing features to create a background: (a) original image of rice grains dispersed for length measurement (note nonuniform illumination); (b) background generated by Gaussian smoothing with 15 pixel radius standard deviation; (c) background generated by replacing each pixel with its darkest neighbor in a 9 pixel neighborhood); (d) subtracting image (c) from image (a) produces uniform background.

the distances between them, and that the background variation is gradual, as it usually is for illumination nonuniformities. In the example of Figure 2.37b this technique is marginally successful, but irregularities in the background remain because the features are too close together for their complete elimination.

A superior method for removing the features to leave just the background uses the rank filter. Rather than averaging the values from the pixels within the features into the background, replacing each pixel with its darkest (or lightest) neighbor will erase light (or dark) features and insert values from the nearby background. This technique uses much the same computation as the median filter, described above, and is another example of ranking pixels in a moving neighborhood. As shown in Figure 2.37c, this method works very well if the features are narrow in at least one direction. It assumes that the features may vary in brightness but are always darker (or lighter) than the local background.

This method can handle rather abrupt changes in brightness, such as can result from shadows, folds in sections, or changes in surface slope. In the example of Figure 2.38, the droplets on a glass surface are illuminated from behind and there is considerable shading, which is not gradual and prevents thresholding to measure their size distribution and spatial arrangement. In this case the contrast around the edges of the droplets is darker than the interior or the background. Replacing each pixel with its brightest neighbor produces a background. Dividing it pixel-by-pixel into the original image produces a image with uniform contrast, which can be measured.

In some cases, it is not sufficient to simply replace pixels by their brighter or darker neighbors, because it can alter the underlying background structure. A more general solution is to first replace the feature pixels with those from the background (e.g., replace dark pixels with lighter ones) and then, when the features are gone, perform the opposite procedure (e.g., replace light pixels with darker ones) using the same neighborhood size. This restores the background structure so that it can be successfully removed, as shown in Figure 2.39. For reasons that will become clear in the next chapters, these procedures are often called erosion and dilation, and the combinations are openings and closings.

Another method that can be used either manually or automatically is to fit a smooth polynomial function to selected background regions dispersed throughout the image. Expressing brightness as a quadratic or cubic function of position in the image requires coefficients that can be determined by least-squares fitting to a number of selected locations. In some cases it is practical to select these by hand, choosing either features or background regions as reference locations that should all have the same brightness value. Figure 2.40 illustrates the result.

If the reference regions are well distributed throughout the image and are either the brightest or darkest pixels present, then they can be located automatically and the polynomial fit performed. The polynomial method requires the variation in brightness to be gradual, but does not require that the features be small or well separated as the rank filtering method does. Figure 2.41 shows an example.

The polynomial fitting technique can be extended to deal with variations in contrast that can result from nonuniform section thickness. Fitting two polynomials, one to the brightest points in the image and a second to the darkest points, provides an envelope that can be used to locally stretch the contrast to a uniform maximum throughout the image as shown in Figure 2.42.

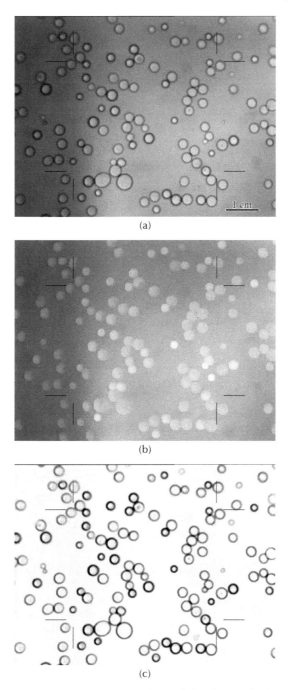

(a)

(b)

(c)

FIGURE 2.38 Removal of a bright background: (a) original image (liquid droplets on glass); (b) background produced by replacing each pixel with its brightest neighbor in a 7-pixel wide neighborhood; (c) dividing image (b) into image (a) produces a leveled image with dark outlines for the droplets.

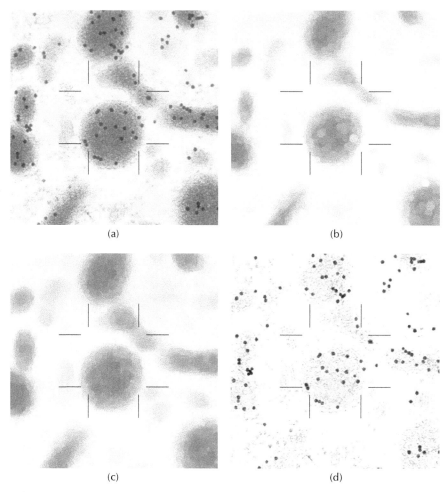

(a)

(b)

(c)

(d)

FIGURE 2.39 Removal of a structured background: (a) original image (gold particles bound to cell organelles); (b) erosion of the dark gold particles also reduces the size of the organelles; (c) dilation restores the organelles to their original dimension; (d) dividing the background from image (c) into the original image (a) produces a leveled imaged in which the gold particles can be counted.

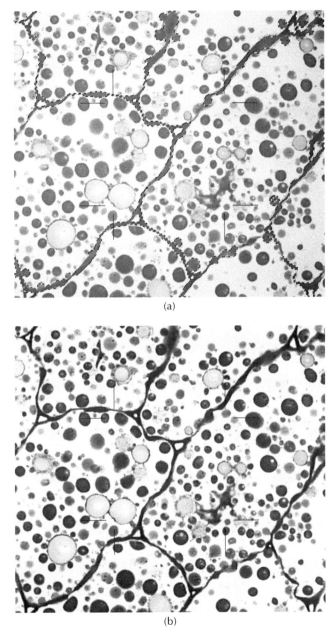

(a)

(b)

FIGURE 2.40 Selecting background points: (a) original image with nonuniform illumination (2-mm-thick peanut section stained with toluidine blue to mark protein bodies, see Color Figure 1.12; see color insert following page 150); the cell walls (marked by the dashed lines) have been manually selected using the Adobe Photoshop® wand tool as features that should all have the same brightness; (b) removing the polynomial fit to the pixels in the selected area from the entire image levels the contrast.

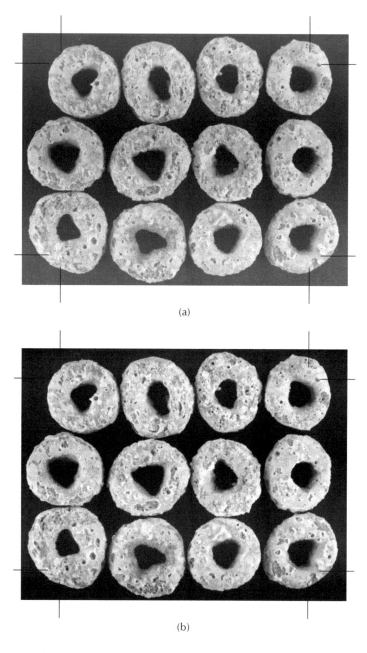

(a)

(b)

FIGURE 2.41 Automatic background correction: (a) original image (courtesy of Diana Kittleson, General Mills) with nonuniform illumination (the blue channel was selected with the best detail from an image originally acquired in color); (b) automatically leveled by removing a polynomial fit to the darkest pixel values.

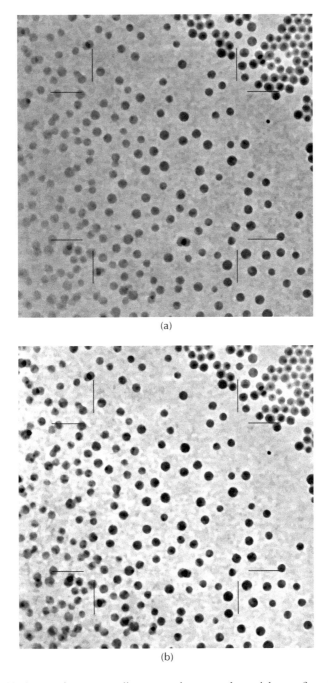

(a)

(b)

FIGURE 2.42 Automatic contrast adjustment using two polynomials, one fit to the dark and one to the light pixels: (a) original (TEM image of latex spheres); (b) processed.

IMAGE DISTORTION AND FOCUS

Another assumption in subsequent chapters on processing and measurement is that dimensions in an image are uniform in all locations and directions. When images are acquired from digital or film cameras on light microscopes, or from scanners, that condition is generally met. Digitized video images often suffer from non-square pixels because different crystal oscillators control the timing of the camera and the digitizer. Atomic force microscopes compound problems of different scales in the slow- and fast-scan direction with nonlinearities due to creep in the piezo drives. Macro cameras may have lens distortions (either pincushion or barrel distortion; short focal length lenses are particularly prone to fish-eye distortion) that can create difficulties.

Electron microscopes have a large depth-of-field that allows in-focus imaging of tilted specimens. This produces a trapezoidal distortion that alters the shape and size of features at different locations in the image. Photography with cameras (either film or digital) under situations that view structures at an angle produces the same foreshortening difficulties. If some known fiducial marks are present, they can be used to rectify the image as shown in Figure 2.43, but this correction applies only to features on the corrected surface.

Distortions in images become particularly noticeable when multiple images of large specimens are acquired and an attempt is made to tile them together to produce a single mosaic image. This may be desired because, as mentioned in the discussion of image resolution, acquiring an image with a large number of pixels allows measurement of features covering a large size range. But even small changes in magnification or alignment make it very difficult to fit the pieces together.

This should not be confused with software that constructs panoramas from a series of images taken with a camera on a tripod. Automatic matching of features along the image edges, and sometimes knowledge about the lens focal length and distortion, is used to distort the images gradually to create seamless joins. Such images are intended for visual effect, not measurement or photogrammetry, and consequently the distortions are acceptable. There are a few cases in which tiling of mosaics is used for image analysis, but in most cases this is accomplished by having automatic microscope stages of sufficient precision that translation produces correct alignment, and the individual images are not distorted.

Light microscope images typically do not have foreshortening distortion because the depth of field of high magnification optics is small. It is possible, however, to acquire a high magnification light microscope image of a sample that has a large amount of relief. No single image can capture the entire depth of the sample, but if a series of pictures is taken that cover the full z-depth of the sample, they can be combined to produce a single extended-focus image as shown in Figure 2.44.

The automatic combining of multiple images to keep the best-focused pixel value at each location requires that the images be aligned and at the same magnification, which makes it ideal for microscope applications but much more difficult to use with macro photography (shift the camera, do not refocus the lens). The software uses the variance of the pixel values in a small neighborhood at each location to

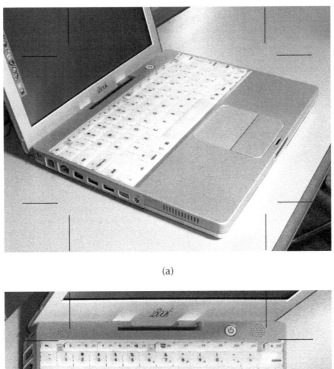

(a)

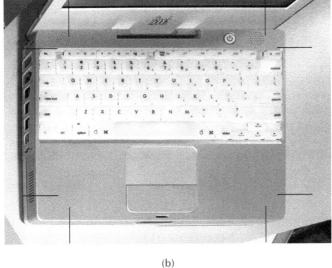

(b)

FIGURE 2.43 Example of correction for trapezoidal distortion: (a) original foreshortened image; (b) calculated rectified view; note that the screen and the edges of the computer, which do not lie in the plane of the keyboard, are still distorted.

select the best focus. For color images, this can be done separately for the red, green, and blue channels to deal with slight focus differences as a function of wavelength.

When an entire image is imperfectly focused, it is sometimes practical to correct the problem by deconvolution. This operation is performed using the Fourier transform of the image, and also requires the Fourier transform of the point spread function (psf). This is the image of a perfect point with the same out-of-focus

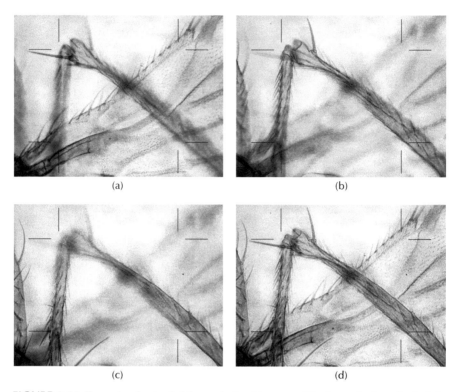

(a) (b)

(c) (d)

FIGURE 2.44 Example of extended focus: (a, b, c) images of the leg and wing of a fruit fly at different focal depths; (d) combination keeping the in-focus detail from each image.

condition as the image of interest. In a few cases it is practical to measure it directly, as shown in Figure 2.45. In astronomy, an image of a single star taken through the same optics becomes the point spread function. In fluorescence microscopy, a single microbead may serve the same function. It is slightly more complicated to obtain the psf by obtaining a blurred image of a known structure or test pattern, deconvolve it with the ideal image to obtain the psf, and then use that to deconvolve other images. That technique is used with the AFM to characterize the shape of the tip (which causes blurs and distortion in the image).

In many cases the most straightforward way to perform deconvolution is to estimate the point spread function, either interactively or iteratively. The two most common types of blur are defocus blur in which the optics are imperfectly adjusted, and motion blur in which the camera or sample are moving, usually in a straight line (Figure 2.46). These produce point spread functions that are adequately modeled in most cases by a Gaussian or by a line, respectively. Adjusting the standard deviation of the Gaussian or the length and orientation of the line produces the point spread function and the deconvolution of the image.

Deconvolution is hampered by the presence of random noise in the image. It is very important to start with the lowest noise, highest bit depth image possible. The noise in the deconvolved result will be much greater than in the original because

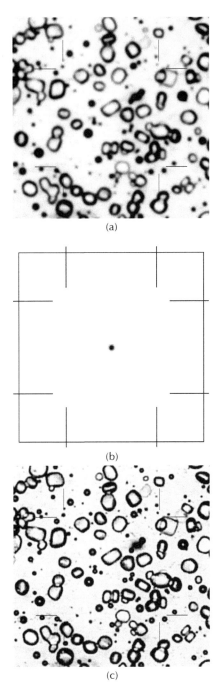

FIGURE 2.45 Deconvolution of an out-of-focus image (ice crystals in ice cream; courtesy of Diana Kittleson, General Mills): (a) original; (b) point spread function measured by using the blurred image of a single small particle; (c) deconvolved result.

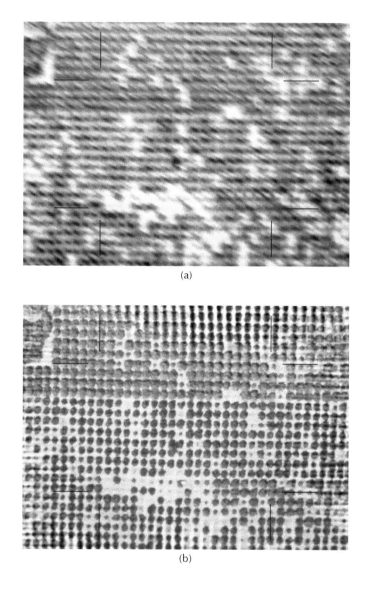

(a)

(b)

FIGURE 2.46 Removal of motion blur: (a) original (aerial photograph of an orchard); (b) deconvolved result using a line with length and direction corresponding to plane motion as the point spread function.

the Fourier transform of the image is divided by that of the psf (this is complex division, because the Fourier transform consists of complex number values). One way to deal with this is to perform a Wiener deconvolution in which a constant is added to the divisor. The constant is a measure of the mean noise level in the original image, but in practice is usually adjusted by the user for an acceptable tradeoff between visible noise and sharpness as shown in Figure 2.47.

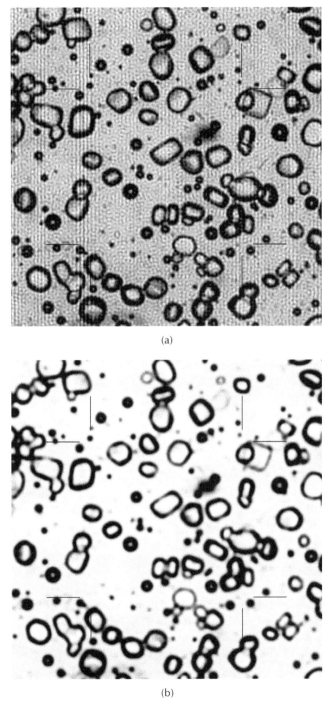

(a)

(b)

FIGURE 2.47 Effect of Wiener constant on deconvolution (same image as Figure 2.45): (a) constant too low, excessive noise; (b) constant too high, too much remaining blur.

Deconvolution is not a routine operation, because in most cases it is easier to retake a blurred image with proper focus. But when the only available picture has some blur, it can be an effective tool to improve the picture quality.

SUMMARY

Image acquisition captures in digital form in the computer an array of pixels that represent the structure and features that will subsequently be measured to characterize the food or food product of interest. There are many different devices that can be used for this image capture, ranging from digital cameras or scanners to instruments such as scanning microscopes that produce the digitized values directly. It is important to understand the specifications and limitations of each. The artifacts and defects in the as-acquired image arise in significant degree from the characteristics of the acquisition device used. Once captured, the images must be saved in a file format that does not degrade the data by lossy compression.

Various kinds of corrections should be considered as part of the acquisition process. Adjustments to correct the color fidelity usually require standards with known colors. Color manipulation of images can be performed using a wide variety of coordinate systems for the three-dimensional color space, each with some particular advantages. For most image processing purposes, a system based on hue, saturation, and intensity is more useful than red, green, and blue because it more closely corresponds to how people perceive color.

Global manipulation of the pixel values is best understood by using the histogram as a guide. Maximizing contrast, either for the entire image or for selected portions of the brightness range, can improve the visibility of structures of interest.

Removal of various kinds of noise from images may be required. Random speckle noise and shot noise are best dealt with by median filters, of which there are several variants that preserve edge and corner contrast and location. Periodic noise is best removed by using a Fourier transform that separates components of the image according to frequency.

Nonuniform brightness in images makes subsequent thresholding of features difficult. It can be corrected by subtracting or dividing by a background image of a uniform scene with the same illumination. If one is not available it can often be generated by processing of the image itself to remove the features, or by fitting polynomial functions to locations selected manually or automatically.

Problems of image distortion can be corrected in some cases by measurement of the distortion or from knowledge of the optics. Limited depth of field can be overcome by combining multiple images that cover the full depth of the subject. Out-of-focus or motion blur can be removed by deconvolution with a measured or estimated point spread function.

The goal of all these procedures is to correct the defects that arise in the image acquisition process so that the best possible image is provided to the next step in the image analysis sequence, which is image processing in order to enhance the details of interest as outlined in the next chapter.

3 Image Enhancement

The previous chapter introduced a variety of methods for processing images that were used to correct for various defects in the as-acquired image, whether those originated in the specimen or the imaging procedure. Related techniques can be used to enhance the visibility or measurability of details in images, and will be discussed in this chapter. Enhancement is generally done for one of two reasons:

1. To improve the visibility of the important details in preparation for printing or otherwise disseminating pictures
2. To isolate the important details from the background to facilitate their measurement

In all cases, it should be understood that enhancement is a two-edged sword. Making the details of interest more visible is accomplished by making other information in the image less visible. This implies a decision about which details are important, and that depends on the context in which the image is being used. As a very simple example, the random speckle noise within uniform features in an image, which can be reduced or eliminated using the methods described in the preceding chapter, is undesirable for purposes of viewing the image, and measuring the dimensions or mean intensity of the region. Eliminating it would be considered a form of enhancement. On the other hand, the amplitude of the noise contains information about the imaging process, the illumination intensity and the camera response. That information might be important if it was needed to verify the conditions under which the image was acquired, or to compare that image to others purported to be similar. Eliminating the noise eliminates the information. Similarly, most processes that alter pixel values to make the differences within the image more visible also destroy any calibration relationship between the pixel values and sample density.

The careful researcher will be aware that subsequent events may indicate the need for different information than was originally extracted from an image, and make it necessary to go back and process it differently to isolate the new data. For that reason, it is always wise to preserve the original image as well as keep track of the processing history of various copies of it. Storage of images is a subject not covered in this text, but one of considerable importance. Obviously, modern computer systems with CD or DVD writers can easily archive images. But the real problem is designing a filing system that makes it possible to find the images again. It is like having to look through all of the photo albums under the bed for that picture of Aunt Hazel in the red dress. If the albums are chronological, like a typical image filing system, you have to hope you remember when it was taken.

Searching computer image databases for key words presupposes that you have saved those descriptors in the first place, and that all of the researchers involved have used the same meaningful and consistent set of words. There are a few software packages that perform searches based on example images: "Find my other pictures that look like this one." At the present level of artificial intelligence, these do not work very well for images of realistic complexity. The most successful application so far has been in searching files of images of paintings, in which they can apparently recognize the styles of some artists.

Returning to the topic of this section, image enhancement, it may be useful to lay out the principal categories of tools that are available. As noted above, most of them have already been illustrated in the preceding chapter.

Global procedures operate on the entire image in the same way. The final value of a pixel is determined by its original value and that value does not vary depending on the local neighborhood. Examples include histogram modification (contrast enhancement, equalization, gamma adjustment, etc.), color correction (and other color space or color channel manipulations), and arithmetic operations that combine multiple images (subtraction, addition, etc.).

Fourier-space procedures convert the image to a different representation, based on the amplitude and phase of the sinusoids that combine to produce it. In this space, filtering can be performed to remove certain frequencies or orientations. Like the global operations, these affect the entire image, but they also depend on the entire image contents. Similar operations can be performed using other transforms such as wavelets, but Fourier techniques are the more familiar and widely used.

Local, or neighborhood operations, consider each pixel in the context of its local neighbors. One class of operations uses kernels of weights to multiply by the pixel values and adds up the resulting total to produce a new pixel. Another class of operators performs statistical calculations with the pixels in the local region. A third class uses ranking of the pixel values to select the median, brightest or darkest. All of these methods operate on every pixel in the image, one at a time, and produce results that alter pixel values differently depending on the values of nearby pixels.

Achieving mastery of image processing tools is primarily developing the experience and ability to understand what each of these types of procedures can do to an image. If you can look at an image and visualize what the effect of a particular technique would be, then you will very quickly be able to choose the method that is most appropriate in any given instance. That is what a skilled and experienced professional does in any field. A journeyman carpenter has the same modest set of tools — hammer, saw, file, screwdriver, etc. — that anyone can purchase at the local hardware store. But he has handled them enough to know exactly what they can do, and knows how to use them to build a house, or a boat, or a piece of furniture. The tools are the same, the experience is the key to using them in the proper way and correct sequence to accomplish the task. Someone who actually does image processing regularly — even a few hours a week — can develop the skills and experience, but it can not be achieved just by reading a book (not even this one). As Nike advertises, you have to "just do it."

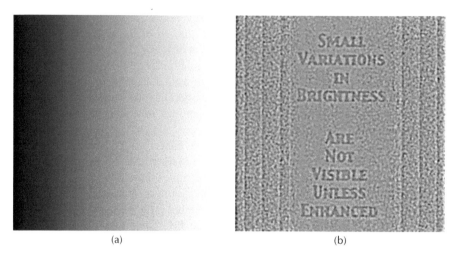

(a)   (b)

FIGURE 3.1 Example of a brightness ramp with superimposed low contrast text: (a) original; (b) processed to show local contrast.

IMPROVING LOCAL CONTRAST

In many images, even though the overall range of brightness values may cover the full dynamic range from white to black, the local contrast that enables detail to be detected may be very faint. Global manipulation of the histogram as discussed in the preceding chapter usually can not correct this, because the local brightness differences are superimposed on the large-scale gradual variations. Figure 3.1 shows an extreme example in which text has been superimposed on a brightness ramp. The letters have only about 1% contrast to the local background, and so are not visible in the original image. Local processing can, however, suppress the global differences and reveal the detail.

The most widely used approach to this task is local equalization. The pixel values in a small, ideally circular neighborhood are used to construct a histogram, and histogram equalization is calculated as described in the preceding chapter. However, the new pixel value is only stored (in a new image) for the central pixel in the neighborhood. The result is that a pixel that was originally slightly darker than its neighborhood becomes darker still, while one that was slightly brighter becomes brighter still. This increases the local contrast (while suppressing large scale variations). It is necessary that the size of the neighborhood be adjusted to the scale of the detail to be enhanced, so that both the dark and light regions fit inside the neighborhood.

Classical local equalization has several problems. It tends to increase the visibility of noise, which must be removed first (in fact, the general rule holds that the procedures for correcting image problems described in the preceding chapter should always be done before attempting enhancement). The sensitivity to noise can sometimes be overcome by using an adaptive or adjustable neighborhood (Figure 3.2) that excludes

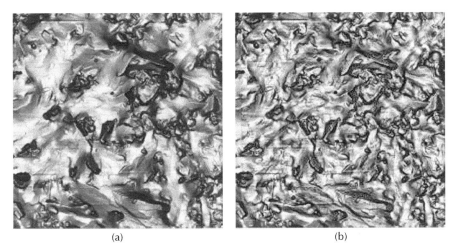

(a)   (b)

FIGURE 3.2 An AFM image of the surface of chocolate showing bloom: (a) original; (b) adaptive equalization. (Courtesy of Greg Ziegler, Penn State University, Department of Food Science)

some pixels based on their extreme values or differences from the central pixel. Second, by eliminating the overall large-scale contrast in the image, local equalization alters the visual appearance so much that landmarks for recognition may be lost. This is typically dealt with by adding back some of the original image to the result.

Finally, when applied to color images it is important that the procedure not be performed on the individual RGB color channels but on the intensity values only, leaving hue and saturation unchanged. As noted before, this is a good general rule for processing color imagery. Processing the color channels directly alters the proportions of red, green and blue and produces distracting color shifts in the results.

There are other related techniques that often perform better than local contrast equalization. One has the confusingly similar name of local variance equalization (Figure 3.3). This also uses a neighborhood that travels across the image, and also alters only the central pixel. But rather than performing histogram equalization on the pixel values, it calculates the local variance in the region and adjusts the values up or down in brightness to make the variance constant across the image. The results tend to be less sensitive to random speckle noise, and less selective in the dimension of the detail that is enhanced.

Another superior technique is based on the retinex algorithm first introduced in the 1960s by Edwin Land, of Polaroid fame. Land studied the characteristics of the human visual system to understand how contrast is perceived, and proposed a method, later implemented in software, for enhancing detail contrast. A more efficient implementation that combines a series of unsharp mask filters (described

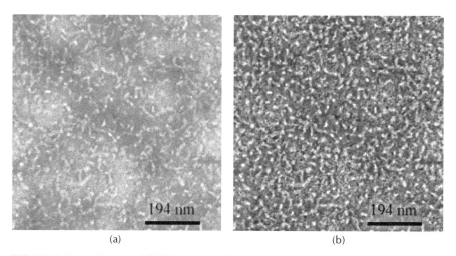

(a)   (b)

FIGURE 3.3 Ovalbumin gel (TEM micrograph courtesy of Erik van der Linden, Food Physics Group, Wageningen University): (a) original; (b) variance equalization.

below) of different sizes has been patented by some NASA engineers. But the most efficient technique for carrying out the essential operations of the retinex technique does not use the pixel array at all, but instead performs in the Fourier domain.

For those wishing a glimpse at the technical details, the procedure works like this: The pixel values are first converted to their logarithms. All of the math is carried out using real numbers, rather than the original integer values for the pixels, but as modern computers handle floating point arithmetic very rapidly, this is no drawback. A Fourier transform is then applied to the log values. In frequency space, an ideal inverse filter is applied. This reduces the amplitude of low frequencies, keeping high frequencies (the filter magnitude is linearly proportional to frequency; this is the same filter that is used in reconstructing images from computed tomography scanners). The filtered data are then inverse Fourier transformed and converted back from the log values with an exponential. The result progressively reduces low frequency (large scale) variations in contrast while enhancing local detail. The conversion to logarithmic values is very important and corresponds to the way that human vision responds to brightness differences.

Thus implemented, retinex processing is a reasonably fast and very effective tool for improving the visibility of local contrast and detail by suppressing but not entirely eliminating large scale contrast, as shown in Figure 3.4. It is particularly effective on images that have a large contrast range with detail in both bright and dark regions, and hence is of great use when dealing with images having more than 256 brightness level (8 bit) dynamic range.

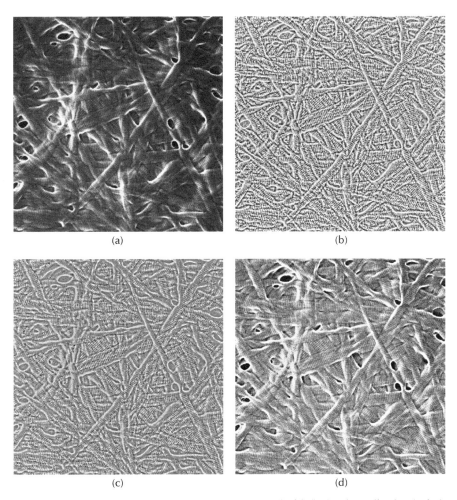

FIGURE 3.4 SEM image of collagen fibers (a) processed with (b) local equalization (noisy); (c) variance equalization (better); (d) retinex (best).

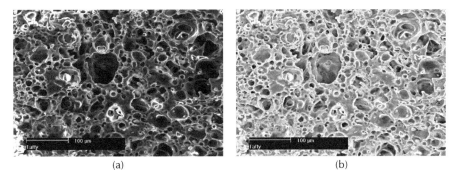

FIGURE 3.5 SEM image of taffy (courtesy of Greg Ziegler, Penn State University, Department of Food Science): (a) original; (b) processed to show details in shadow regions.

IMAGE SHARPENING

The visual appearance of image detail is improved by increasing the amount of contrast at well defined edges and steps. This result is often accomplished by applying high pass filters to the data, meaning that high frequencies are kept or increased in amplitude while low frequencies are reduced or eliminated. The terminology corresponds to performing the processing in Fourier space, but in practice the operations are often performed directly on the pixel array by the use of convolution kernels — arrays of weights that are multiplied by the pixel values and the results summed to produce a new value for the pixel. Kernels were introduced in the preceding chapter as a way to perform smoothing, which is a low pass filter that reduces high frequencies — in that case random speckle noise.

To better understand how these filters work it is instructive to start with a simple derivative. Unlike the smoothing filters shown previously, the weight values in these kernels are neither symmetric about the center nor all positive. A representative first derivative kernel is an array of the form

+1	+2	+1		+1	+2	+4	+2	+1
0	0	0	or	0	0	0	0	0
−1	−2	−1		−1	−2	−4	−2	−1

The operation of this kernel is to replace the value of each pixel with the difference between the neighbors above and below, hence forming a vertical derivative. The weights perform some averaging in the horizontal direction to reduce random noise. Because of the negative weights, it is possible for the result of applying this kernel to be negative. Since the display can usually only show values in the range 0 (black) to 255 (white), the result is usually divided by the largest number in the array (2 or 4 in the examples) and has a medium grey value of 128 added to it so that negative or positive results are visible. (A few programs perform the additional step of autoscaling the result to maximize contrast without clipping.) Figure 3.6 shows an example. The effect of the derivative is to produce an embossed appearance in the image, in which horizontal edges have a bright and a dark edge. Vertical detail is not enhanced, and in fact virtually disappears.

A directional derivative may be useful for eliminating the visual distraction produced by strong shadows, scratches, or other directional information superimposed on an image. In the example shown in Figure 3.7, the grains in the surface of aluminum foil (used as food packaging) are obscured by the linear marks of the extrusion or rolling process. Applying a directional derivative exactly parallel to the marks makes them disappear so that the underlying structure is easier to examine.

While these embossed images may be visually attractive, they are only rarely useful for measurement purposes because of the directionality of the enhancement. A second derivative can be similarly constructed which is non-directional. A kernel of the form

				0	−1	−2	−1	0
−1	−1	−1		−1	−1	+3	−1	−1
−1	+8	−1	or	−2	+3	+8	+3	−2
−1	−1	−1		−1	−1	+3	−1	−1
				0	−1	−2	−1	0

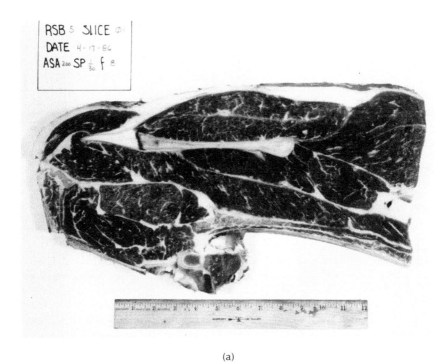

(a)

(b)

FIGURE 3.6 Image of a beef roast (a) and the result of applying a vertical derivative (b).

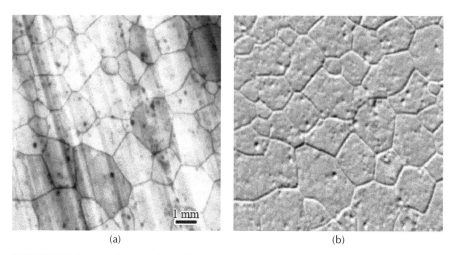

(a)   (b)

FIGURE 3.7 Image of the surface of an aluminum foil (a) and the result of application of an embossing (derivative) filter parallel to the extrusion or rolling lines to hide them (b).

is a classic Laplacian operator. It is necessary to add an offset value to allow negative results to be shown for the derivative. By taking the difference between a central pixel (or a small central region) and its surroundings, the Laplacian is sensitive to brightness changes that occur in any direction. The result, as shown in Figure 3.8, is to keep only the changes in brightness and to mark each one with a black and a white edge.

The problem is the same as for local equalization in that the overall contrast is lost, but this can be restored by adding the result to the original image. The usual short cut for that operation is to add 1 to the weight for the central pixel (and eliminate the offset value). That produces the most widely used sharpening filters, which increase the amplitude of brightness changes without altering the overall contrast as shown schematically in Figure 3.9. This sharpening operation is commonly applied to photographs before printing them, because most printing operations cause a certain amount of blurring, which the filtering corrects.

These sharpening filters do not actually sharpen the image in the sense of correcting out-of-focus blur, as described in the preceding chapter. What they do is trick the eye by increasing contrast where it already exists. This is almost entirely used for improving the visual appearance of images, and not for facilitating any subsequent measurement or analysis of the images.

The sharpening filter shown above has two important limitations: it is sensitive to speckle noise (a pixel that is slightly different from its neighbors has that difference increased) and it only increases the contrast for pixels that differ from their immediate neighbors. A more general technique for sharpening is the so-called "unsharp mask" filter, which is itself a specialized version of the most general form of this approach, the difference-of-Gaussians (DoG) filter. This latter procedure is believed to be similar to the processing performed by neural networks in the human retina, and can also be shown to correspond to a band-pass filter performed using Fourier transforms.

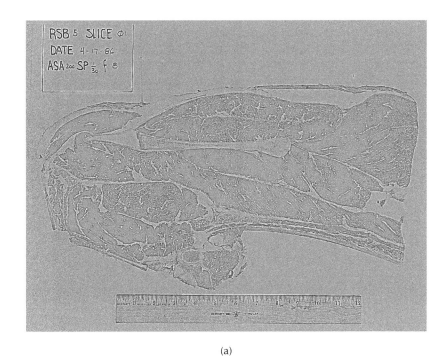

(a)

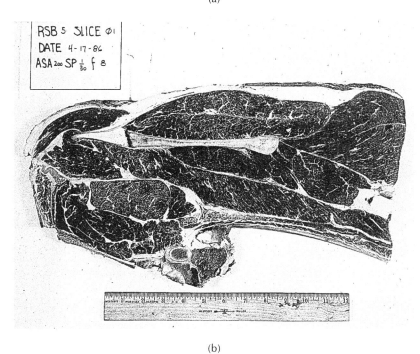

(b)

FIGURE 3.8 The beef image from Figure 3.6 after application of (a) Laplacian and (b) sharpening filters.

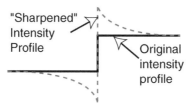

"Sharpened" Intensity Profile

Original intensity profile

FIGURE 3.9 Diagram showing the increase in contrast at edges produced by sharpening operations.

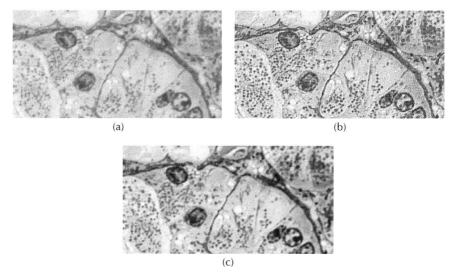

(a)

(b)

(c)

FIGURE 3.10 Light micrograph of stained tissue (the original image is in color, but the process is applied only to the intensity or luminance channel): (a) original (b) Laplacian; (c) DoG.

The DoG method (Figure 3.10) uses two copies of the original image. One is blurred with a Gaussian filter having a very small standard deviation, typically ranging from less than 1 to a few pixels, in order to reduce random pixel variations (noise). The second is also blurred with a Gaussian filter with a standard deviation about 3 to 5 times larger. The larger blur is intended to selectively remove important edges and detail from the image. The difference between the two images isolates those important edges and detail. As for the other sharpening and enhancement methods, these details are then often added back to the original image for viewing. This general approach of finding a way to remove the important information from an image and then recovering it by subtraction from the original is a useful technique in many situations.

In the particular case in which the original image does not contain much random noise, the first (small) blur is unnecessary and the DoG method reduces to the unsharp mask. In this case a blurred copy of the image is subtracted from the original. The name originates in the photographic darkroom, describing a century-old practice.

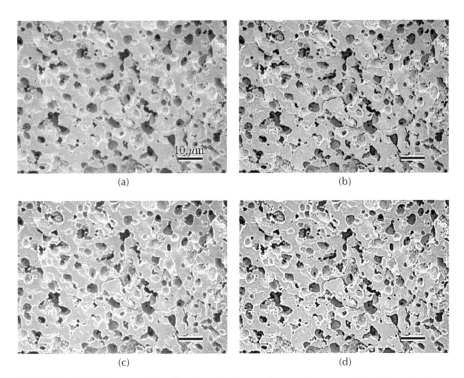

FIGURE 3.11 Unsharp mask or DoG applied to an image of a cut surface through cheese (image courtesy of A. J. Pastorino, Department of Nutrition and Food Sciences, Utah State University): (a) original; (b) adding only the dark edge contrast; (c) only the light edge contrast; (d) both.

Photographic film has a much greater latitude than paper prints, and a negative can therefore capture bright and dark portions of a scene that are difficult to show in the final print. One solution is to first use the negative to expose a second piece of film, at a 1:1 scale but slightly out of focus. Developed, this is the unsharp (because it is out of focus) mask. Placing the two pieces of film together and printing through both produces reduced overall contrast, because the mask is dark where the original negative is light and vice versa, but preserves edge detail because the mask is not sharp. The computer procedure corresponds to that of the darkroom, but with the added convenience of interactively controlling the amount of blur and the amount of subtraction.

The effect of a DoG or unsharp mask filter is to increase the magnitude of contrast at edges and other fine detail, as indicated in Figure 3.9. Some programs allow selectively keeping the added contrast on either the bright or dark side of the edge, which can also further improve the visual appearance of the result (Figure 3.11).

The DoG and other sharpening filters are generally described as "hi-pass" filters because in terms of the Fourier transform and the frequency-space representation of the image, they reduce the amplitude of low frequency terms that control global contrast while keeping the higher frequency terms that produce local detail. The DoG refines this approach further by reducing the very highest frequency terms

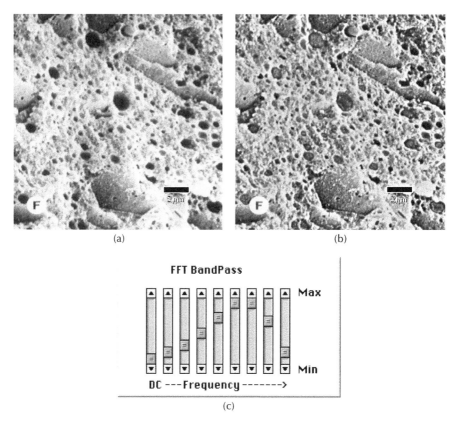

(a)   (b)

(c)

FIGURE 3.12 SEM image of processed Colby cheese (a) the original image appears in Aguilera and Stanley, Figure 4-2f, courtesy of the authors and Ken Baker, Ken Baker Associates and (b) the result of applying a bandpass filter. (c) Each slider in the equalizer adjusts the amplitude of frequencies higher by a factor of 2.

(corresponding to the random pixel noise). That suggests that the greatest control over detail enhancement can be effected by working directly with the Fourier transform. It is possible to construct a filter that keeps whatever range of frequencies corresponds to the most interesting image detail while suppressing others that represent noise or unwanted global contrast, much in the manner of adjusting an audio equalizer to get the best sound from a given set of speakers. Figure 3.12 shows that this method can enhance detail and at the same time suppress unwanted contrast so that information can be recovered from deeply shadowed pits in an SEM image.

This type of range compression is particularly useful for images in which different regions are very bright or very dark. As noted previously, human vision detects local changes in brightness of a few percent, and largely ignores large scale differences in brightness. A method that compresses the large scale variations while increasing the local differences makes details in bright or shadow areas more visible. SEM images of rough surfaces, such as the example in Figure 3.5, can often be improved by this technique.

FALSE COLOR AND SURFACE RENDERING

The reason that sharpening methods are useful to visually enhance detail in images lies in the limitations of human vision, which does not notice gradual changes in brightness nor variations of less than a few percent. Another very popular way to make small differences in brightness visible is to assign false colors to the image. For an 8 bit image with 256 stored brightness levels, it is straightforward to construct a table of 256 colors (usually called a "CLUT" or "Color Look Up Table") and assign one to each of the possible brightness levels. Human vision can discern only about 20 to 30 brightness levels in an image, but can distinguish hundreds of colors, so every different pixel value can be readily observed.

Because it emphasizes the difference between pixels, the application of false color (pseudocolor) does not help the observer to see the relationship between features in an image. False color images are very good for discerning gradual changes in brightness in large, nearly uniform areas. Applied to a typical image, they may act much like camouflage and actually hide structure by breaking it up. Applying a CLUT in which the colors vary gradually, for instance in a spectrum or heat scale, causes less break up in the Gestalt of an image but also does not provide as much ability to visually distinguish small changes. These pictures do, however, remain popular with magazine editors who want a splash of color for the front cover.

While human vision is a poor judge of changes in brightness or color, we do have millions of years of evolution to guide the interpretation of surface images. The appearance of light interacting with a surface, particularly if the angle of incidence of the illumination and the angle of viewing can be altered, is instinctively interpreted to provide understanding of the surface geometry and roughness, color, and reflectivity. The mathematics of light scattering and reflecting from surfaces is well understood and computer programs are routinely used in CAD applications, movie animations, and advertising to model the appearances of physical surfaces. Using the same math to generate a rendering of an image as though it were a surface often reveals details that were difficult or impossible to see in the original image.

There are two parts to surface displays. One is to create a geometric representation of a three-dimensional surface (or something we are pretending is a surface) on a two-dimensional computer screen. Geometrical modeling produces a perspective-corrected view from a chosen point in space in which locations are displaced upwards according to some value (usually pixel brightness) to represent the surface. Surface rendering then controls the brightness of that point based on the local angle of the surface facets, the intensity and position of the incident light, and the surface characteristics, specifically how much of the light is specularly reflected from the surface and how much of it is diffusely scattered. Figure 3.13 illustrates these possibilities for enhancing the visibility and interpretability of detail.

Many atomic force microscopes offer at least geometric modeling, and some include surface rendering of the image, to show their images. In many cases, the geometry revealed in the images is literally the elevation as a function of position (Figure 3.14), which is one of the major measurements that the microscope provides. Since AFM instruments can also record many other characteristics of the sample-tip interaction (e.g., lateral force, tapping compliance, heat conductivity, etc.), each

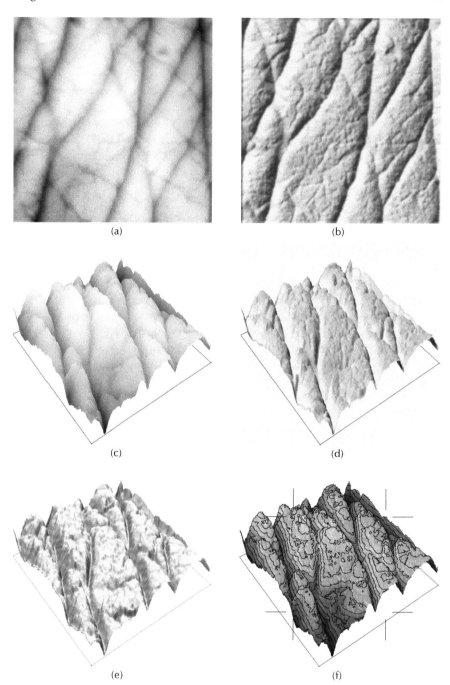

FIGURE 3.13 Scanned probe image of the surface of skin: (a) original (image width 1 cm); (b) the surface Phong rendered; (c) a geometric model constructed from the data; (d) the geometric model with superimposed surface contrast rendered with diffuse lighting; (e) same as (d) but with specular lighting; (f) the surface color coded to indicate elevation, with superimposed contour lines (see color insert following page 150).

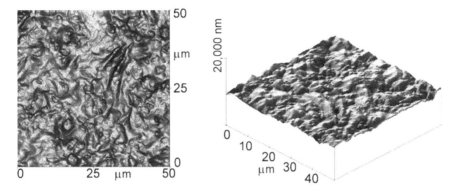

FIGURE 3.14 AFM image of the surface of chocolate with a rendering showing the physical geometry of the actual surface. (Courtesy of A. J. Pastorino, Department of Nutrition and Food Sciences, Utah State University)

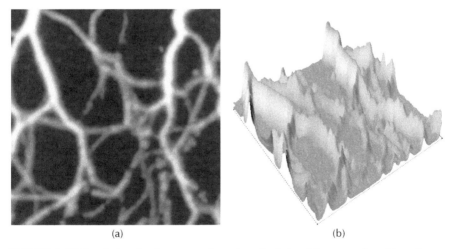

(a)   (b)

FIGURE 3.15 Rendering the fluorescent intensity of tubules (a) as a surface (b), to better see faint marks and compare intensities.

of these can be shown as though the values represented a surface, as well. Adding color and rendering of the surface (usually making the simplifying assumption that the surface is a highly specular reflector) can also be used to help reveal subtle detail to the eye.

As shown in Figure 3.15, presenting grey scale intensity as though it represents a surface is also useful with other types of images. The example shown is a fluorescence image of stained tubules. When the intensity is interpreted as surface elevation and the resulting data rendered, the faint markings may become easier to see and the differences in intensity easier to interpret.

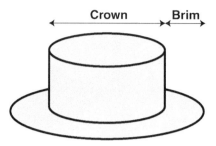

FIGURE 3.16 Schematic drawing of a top-hat filter.

RANK-BASED FILTERS

In the preceding chapter, two very different categories of neighborhood procedures were introduced. One used a convolution kernel of weights that were multiplied by the pixel values, and the results summed, while the other was based on ranking the pixels in the neighborhood into order and selecting the median, brightest or darkest value. The sharpening methods described above are convolution methods that use all of the pixels in the neighborhood, with weights that depend on distance and direction from the central pixel. It is also useful to apply rank-based procedures to select details of interest for enhancement.

The top-hat filter is used to select objects that are darker (or brighter) than the local background for retention or removal. The procedure uses two neighborhoods for ranking. Usually the central one is circular and the outer one is an annulus that surrounds it. The name comes from drawing this configuration (Figure 3.16) with the inner neighborhood shown as the crown of a top hat and the outer one as the brim. The height of the crown is the required brightness difference between the feature (which must fit inside the crown) and the background.

The top hat compares the darkest pixel in the inner region to the darkest one in the surrounding neighborhood. If the difference between these is greater than some threshold value the pixel is kept, otherwise it is not. This allows selection of features based on size (defined by the inner region), contrast (the required difference in pixel values), and separation (the width of the brim). As shown in Figure 3.17, features that are large, or close together, or have long lines, are rejected because the darkest values in the inner and outer neighborhoods are not sufficiently different. Obviously, the logic can be reversed to look for bright features instead of dark ones.

One of the principal uses of the top-hat filter is locating spikes in Fourier-space power spectra. The high frequency periodic noise spikes shown in the preceding chapter are efficiently located by a top-hat filter, rather than by manual marking. A top hat was used in this way to create the filter that removed the periodic noise in Figure 2.35 of the preceding chapter. The procedure is also useful whenever images contain features of interest that have a particular size, especially when the image also contains other larger objects, possibly on a nonuniform background. Figure 3.18 shows an example.

If the top-hat filter can be used to isolate a particular structure from an image, it follows that it can also be used to remove it. When used for that purpose, it is a

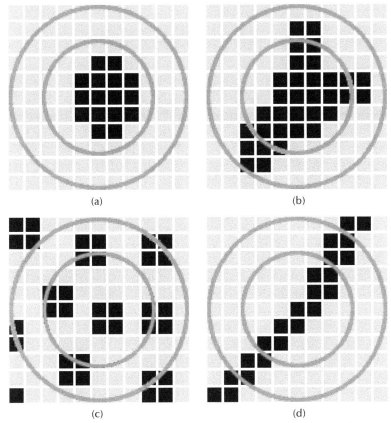

(a)   (b)

(c)   (d)

FIGURE 3.17 Illustration of the operation of a top-hat filter: (a) features that fit entirely inside the crown are found; (b to d) features that do not fit inside the crown, or have neighbors that hit the brim, are ignored.

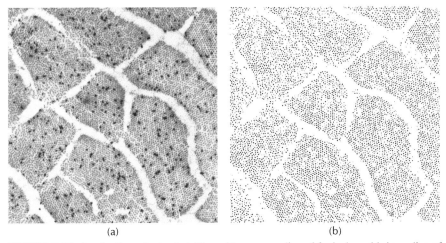

(a)   (b)

FIGURE 3.18 Application of a top-hat filter with crown radius of 2 pixels and brim radius of 4 pixels to locate the stained fibers while ignoring larger features on a cross-section of muscle.

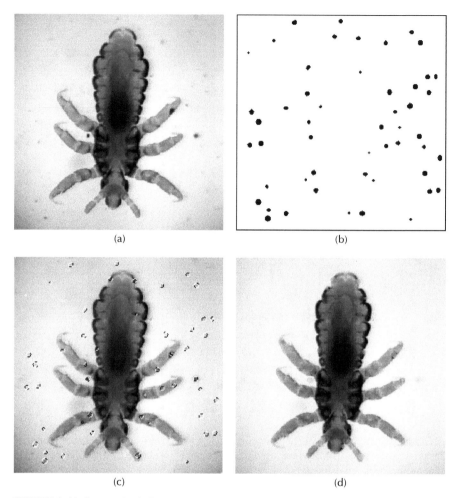

FIGURE 3.19 Removal of dirt: (a) original image showing a bug on a dirty slide; (b) dirt particles located by the top-hat filter; (c) the outlines of the dirt particles selecting regions on the original image; (d) filling the regions in image (c) by interpolation to remove the dust particles.

rolling ball filter. Features that are small enough to fit inside the crown of the hat, and either lighter or darker than the surrounding brim by more than some set threshold, are replaced by pixel values from the brim. In its simplest form this is done by using the average value from the brim, but it is also possible to interpolate to generate a smoother appearance to replace the selected features (Figure 3.19).

EDGE-FINDING

A significant fraction of the image-processing literature is concerned with procedures for locating edges in images, defined as locations where the brightness (or color) changes abruptly and which often correspond to the boundaries of features. This seems to be a function that human vision carries out in the very early stages

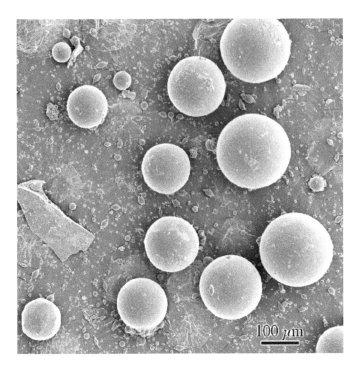

FIGURE 3.20 SEM image of lipid particles.

of processing the raw information from the millions of light-sensing cells in the retina, producing what is often described as a sketch of the scene to be sent to higher levels of the brain. For image measurement purposes, the edges of features are important to define size and shape. Many of the stereological procedures used to characterize global structural properties also use the edges, for instance either to measure their total length or to count the number of intersections they make with a grid of lines.

Edges are also important because many of the types of microscopes used to study the structure of food products produce images in which there is a localized change of brightness associated with the edges of features or structures, even though the interior of the various regions may be very similar in brightness or color. This is often the case with SEM images of particles, for instance, because as illustrated in Figure 3.20, the edges appear bright (highly sloped surfaces emit more electrons) while the centers of the particles have the same slope and the same brightness as the substrate.

Situations arise in which the edges of a feature can be defined by a change in brightness from the background, but the interior of the feature has widely varying brightness or color values. Figure 3.21 shows two different examples of bubbles, one in a fluid and another in a solid. In the liquid, the bubble interiors are bright on one side and dark on the other, but the border can be distinguished by its difference from the local background. In the solid, some of the bubble interiors are bright, some are dark, and some are the same as the matrix.

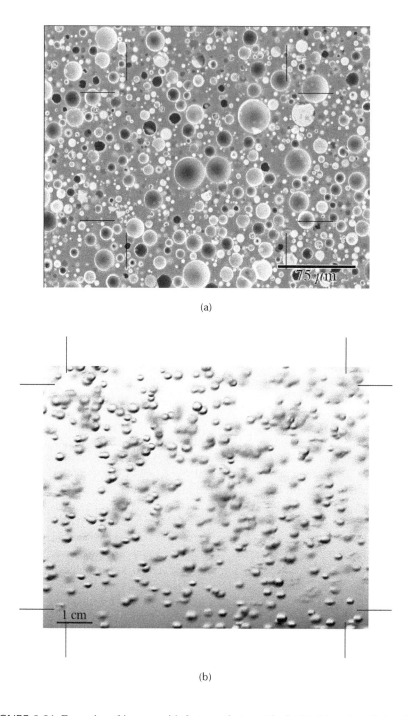

(a)

(b)

FIGURE 3.21 Examples of images with features that must be isolated based on their edges rather than their internal brightness: (a) bubbles in an epoxy resin; (b) bubbles in a liquid, illuminated from the top.

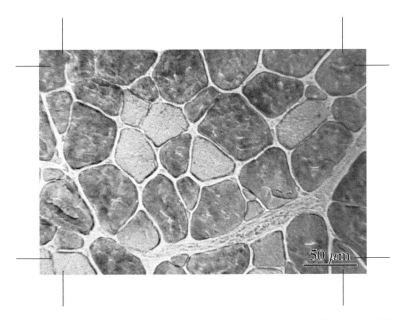

FIGURE 3.22 Transverse section of pork reacted for phosphorylase. (Courtesy of Howard Swatland, Dept. of Animal and Poultry Science, University of Guelph)

A third instance in which edges are important for structural determination occurs when various regions that are functionally similar have different brightness levels, and it is only the change in brightness from one to another that defines the boundaries. Figure 3.22 shows an example in meat.

To deal with these cases, a variety of algorithms have been developed. The most general and consequently the most widely used will be described and illustrated here, but it should be noted that there are many very specialized techniques that rely on prior knowledge about the nature of a particular type of image in order to efficiently and accurately determine the position of boundaries. This is particularly true for medical images where the basic components of structure are well known (the organs in the human body) and the task of software is to fit the expected boundaries onto a particular image, within constraints.

Probably the most widely used edge-finding tool is the Sobel magnitude filter. This calculates the magnitude of the local intensity gradient in the image, by applying two kernels to every pixel location. The directional derivative shown above in Figure 3.6 calculated the vertical change in brightness. The array of kernel values can easily be rotated by 90° to calculate the directional derivative in the horizontal direction. If the two derivative values are considered to be vector magnitudes in the horizontal and vertical directions, then they can be combined to calculate the overall vector as shown in Figure 3.23. The magnitude of this vector is the Sobel result, which is assigned to the pixel. The angle of the vector can be calculated as well, and will be used for other purposes below.

(a)

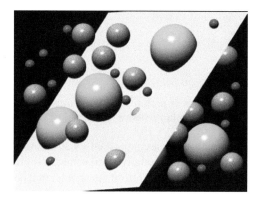

(b)

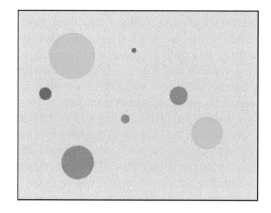

(c)

COLOR FIGURE 1.1

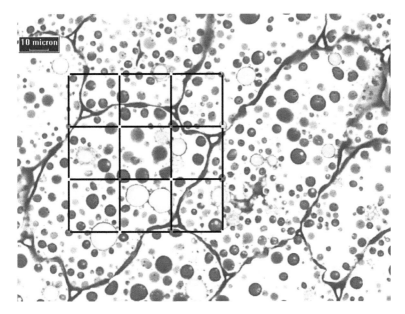

COLOR FIGURE 1.12 The color plate also corresponds to Figures 2.40 and 4.41.

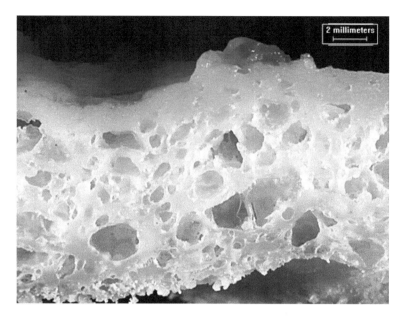

(a)

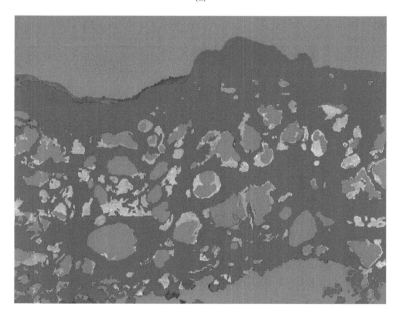

(b)

COLOR FIGURE 1.26

COLOR FIGURE 2.3 The color plate also corresponds to Figures 2.10 and 2.21.

(a)

(b)

COLOR FIGURE 2.11

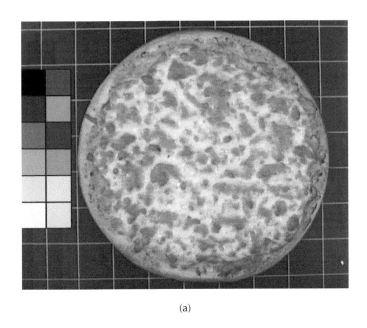

(a)

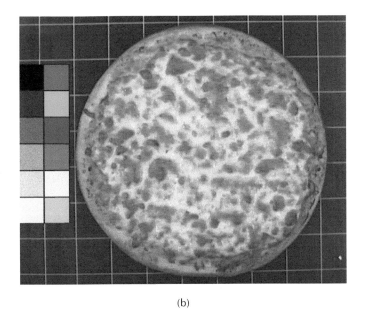

(b)

COLOR FIGURE 2.13

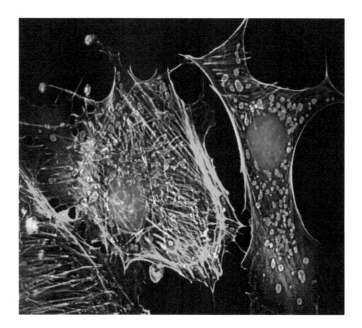

COLOR FIGURE 2.15

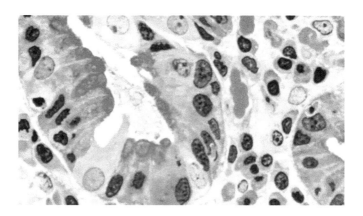

COLOR FIGURE 2.16

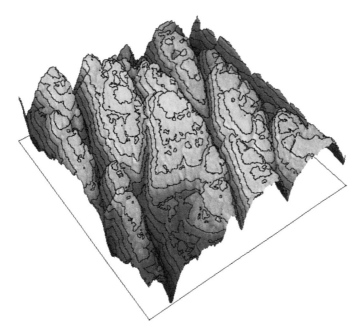

COLOR FIGURE 3.13

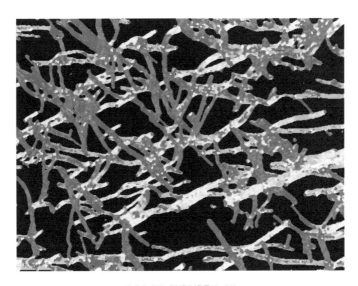

COLOR FIGURE 3.37

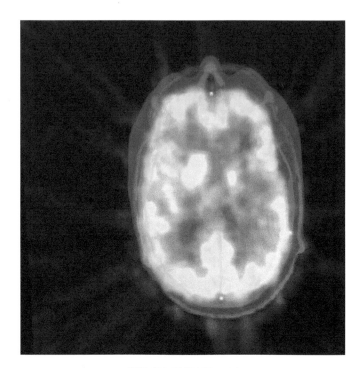

COLOR FIGURE 3.46

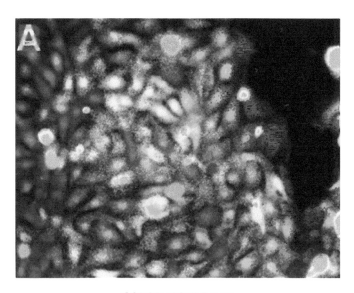

COLOR FIGURE 3.48

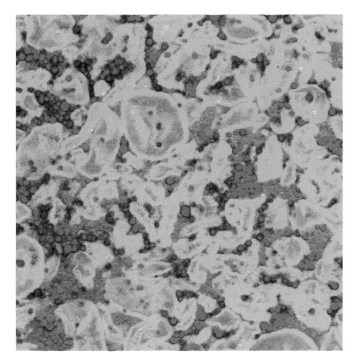

COLOR FIGURE 3.56a The color plate also corresponds to Figure 5.29a.

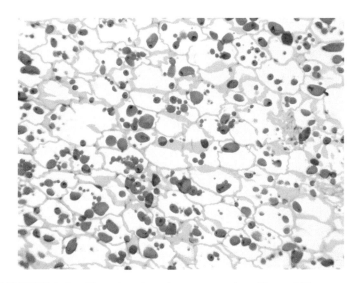

COLOR FIGURE 3.57a The color plate also corresponds to Figure 5.30a.

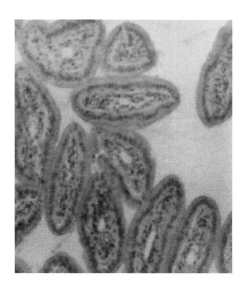

COLOR FIGURE 3.58

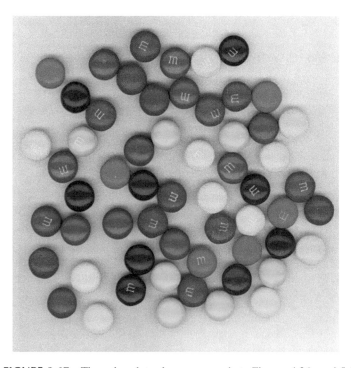

COLOR FIGURE 3.67a The color plate also corresponds to Figures 4.26a and 5.10a.

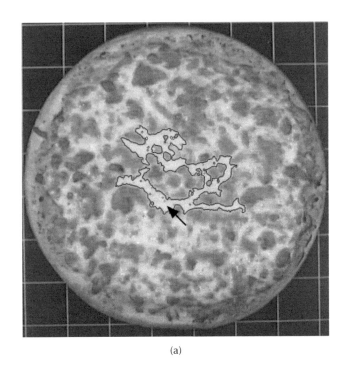

(a)

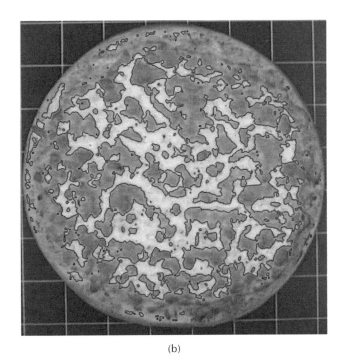

(b)

COLOR FIGURE 3.68

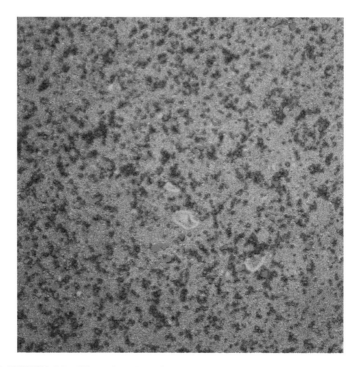

COLOR FIGURE 4.7a The color plate also corresponds to Figure 5.29b.

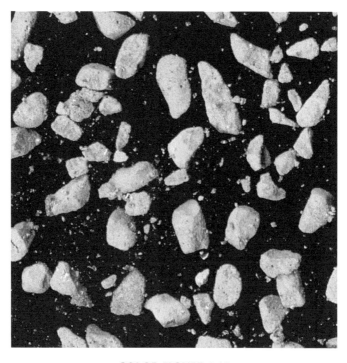

COLOR FIGURE 4.13

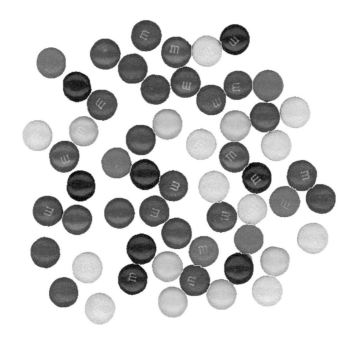

(a)

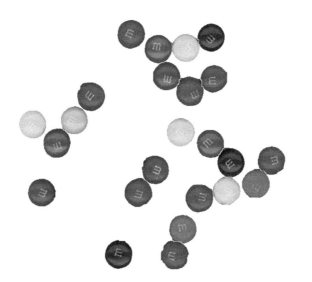

(b)

COLOR FIGURE 4.26

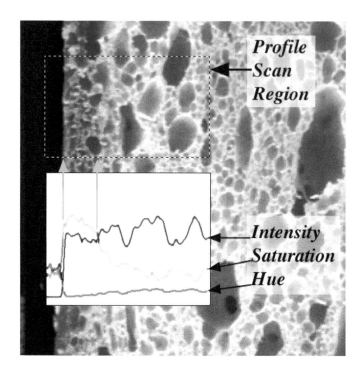

COLOR FIGURE 5.26

COLOR FIGURE 5.52

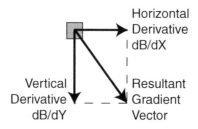

FIGURE 3.23 Calculation of the Sobel filter, the magnitude of the intensity gradient vector.

$$Magnitude = \sqrt{\left(\frac{dB}{dX}\right)^2 + \left(\frac{dB}{dY}\right)^2}$$

$$Angle = \arctan\left(\frac{dB/dY}{dB/dX}\right)$$

(3.1)

The Sobel magnitude filter outlines feature boundaries well and is efficient to compute, but has the disadvantage that it measures the absolute brightness change at the boundary, whereas visually the boundary contrast is perceived as a percentage change. More advanced (and computationally complicated) edge finding operators such as the Frei and Chen filter overcome this problem, and are also better able to distinguish between edges, lines, and noise. Instead of the 2 kernels that the Sobel applies at each pixel location, the Frei and Chen applies 7, and combines the results. Figure 3.24 compares the results of these two operations.

Another modification of the gradient operator thins the boundary lines to single-pixel width to mark the exact location of the edge. The Canny operator follows the kernel or convolution operations with a ranking of the resulting magnitude values that compares each pixel to its neighbors and erases those that have neighbors with greater magnitudes. This leaves just the thinned line along the maximum values and thus marks the most probable edge position (Figure 3.25).

There are many other approaches to edge location, which produce visually similar results on many images but have advantages in certain cases. A very simple way to locate a change in brightness is to calculate the difference in brightness between the maximum and minimum pixel values in a small neighborhood. The brightness range increases when the neighborhood is located over an edge. Another statistical operation calculates the variance of the pixel values in a small neighborhood. This value rises significantly when the neighborhood is located at an edge, and can detect quite subtle boundaries in images. Figure 3.26 illustrates these methods.

All of these edge-finding methods attempt to mark edge locations. These will be used in later sections of this chapter and the next to threshold those outlines for measurement, or to fill in the boundaries to delineate the corresponding features. There is another type of edge finding operation, which does not eliminate the

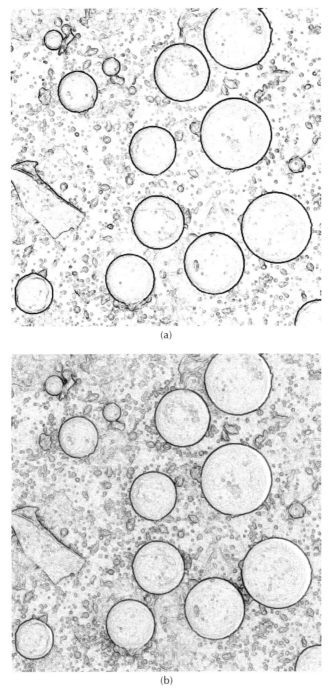

(a)

(b)

FIGURE 3.24 Edge filtering the image from Figure 3.20: (a) Sobel; (b) Frei and Chen.

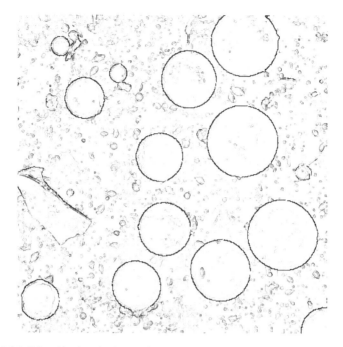

FIGURE 3.25 Edge filtering the image from Figure 3.20 with a Canny operator.

brightness change at the boundary but instead sharpens it. A useful analogy is to the erosion of buttes in the deserts of Utah. An idealized butte or mesa has a flat top and sheer sides that drop away to the flat floor of the desert below. Erosion causes these abrupt changes in elevation to become gradually worn down, until the sides are sloped (more like the shape of hills in the older terrain of the eastern United States). To accurately measure the size of such a feature, it might be useful to reverse the erosion and put the eroded material back on top so that the mesa again had sheer sides.

That type of reconstruction of feature boundaries is performed with a maximum likelihood operator. Each pixel is compared to the various neighborhoods on every side that it might belong to, and a statistical test (e.g., variance and entropy among many possible choices that produce generally similar results) is applied to decide which neighborhood the pixel is most like. The pixel is then given either the mean or median value of the neighborhood to which it is assigned. As shown in Figure 3.27, the result is to assign doubtful pixels to one region or another and to create abrupt transitions that facilitate the measurement of features. One consequence of this procedure is that pixels which straddled a boundary and have intermediate brightness values were represented by a broad, flat valley in the image histogram; but after a maximum likelihood reconstruction, the valley is reduced or eliminated, which simplifies the subsequent thresholding steps discussed below.

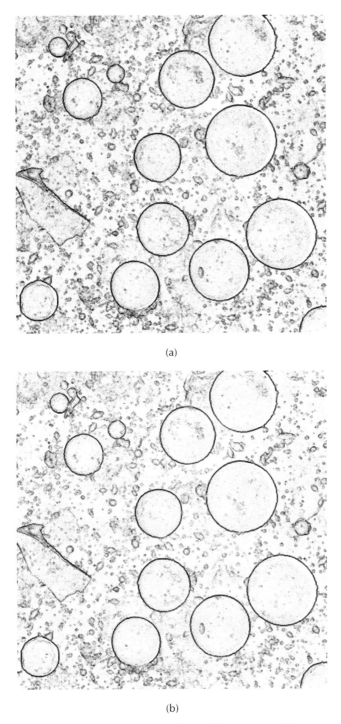

(a)

(b)

FIGURE 3.26 Edge filtering the image from Figure 3.20: (a) range; (b) variance.

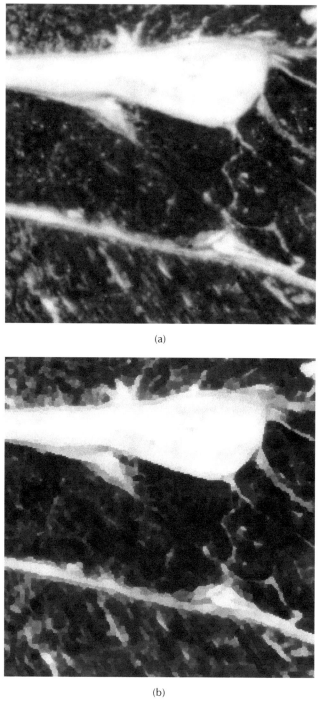

(a)

(b)

FIGURE 3.27 Application of a maximum likelihood filter to the beef image from Figure 3.6(a) (detail enlarged to show individual pixels): (a) original; (b) processed.

TEXTURE

Many structures are not distinguished from each other or from the surrounding background by a difference in brightness or color, nor by a distinct boundary line of separation. Yet visually they may be easily distinguished by a texture. This use of the word "texture" may initially be confusing to food scientists who use it to describe physical and sensory aspects of food; it is used in image analysis as a way to describe the visual appearance of irregularities or variations in the image, which may be related to structure. Texture does not have a simple or unique mathematical definition, but refers in general terms to a characteristic variability in brightness (or color) that may exist at very local scales, or vary in a predictable way with distance or direction. Examples of a few visually different textures are shown in Figure 3.28. These come from a book (P. Brodatz, *Textures: A Photographic Album for Artists and Designers*, Dover Publications, New York, 1966) that presented a variety of visual textures, which have subsequently been widely distributed via the Internet and are used in many image processing examples of texture recognition.

Just as there are many different ways that texture can occur, so there are different image processing tools that respond to it. Once the human viewer has determined that texture is the distinguishing characteristic corresponding to the structural differences that are to be enhanced or measured, selecting the appropriate tool to apply to the image is often a matter of experience or trial-and-error. The goal is typically to convert the textural difference to a difference in brightness that can be thresholded.

Figure 3.29 illustrates one of the simplest texture filters. The original image is a light micrograph of a thin section of fat in cheese, showing considerable variation in brightness from side to side (the result of varying slice thickness). Visually, the smooth areas (fat) are easily distinguished from the highly textured protein network around them, but there is no unique brightness value associated with the fat regions, and they cannot be thresholded to separate them from the background for measurement. The range operator, which was used above as an edge detector, can also be useful for detecting texture. Instead of using a very small neighborhood to localize the edge, a large enough region must be used to encompass the scale of the texture (so that both light and dark regions will be covered). In the example, a neighborhood radius of at least 2.5 pixels is sufficient.

Many of the texture-detecting filters were originally developed for the application of identifying different fields of planted crops in aerial photographs or satellite images, but they work equally well when applied to microscope images. In Figure 3.30 a calculation of the local entropy isolates the mitochondria in a TEM image of liver tissue.

Probably the most calculation-intensive texture operator in widespread use determines the local fractal dimension of the image brightness. This requires, for every pixel in the image, constructing a plot of the range (difference between brightest and darkest pixels) as a function of the radius of the neighborhood, out to a radius of about 7 pixels. Plotted on log-log axes, this often shows a linear increase in contrast with neighborhood size. Performing least-squares regression to determine

FIGURE 3.28 A selection of the Brodatz textures, which are characterized by variations in brightness with distance and direction.

the slope and intercept of the relationship measures the fractal parameters of the image contrast (for a range image, as would be acquired from an AFM, it is the actual surface fractal dimension; for an image in which the brightness is a function of slope, as in the SEM, it is equal to the actual surface dimension plus a constant).

As shown in the example of Figure 3.31, these parameters also convert the original image in which texture is the distinguishing feature to one in which brightness can be used for thresholding. The fractal texture has the advantage that it does not depend on any specific distance scale to encompass the bright and dark pixels that make up the texture. It is also particularly useful because so many structures in the real world are naturally fractal that human vision has apparently evolved to recognize that characteristic, and many visually discerned textures have a fractal mathematical relationship.

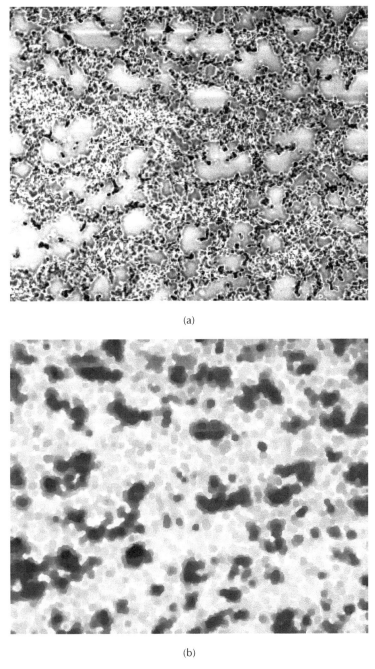

(a)

(b)

FIGURE 3.29 Thresholding the fat in cheese: (a) original image; (b) texture converted to a brightness difference using the range operator; (c) the fat regions outlined (using an automatic thresholding procedure as described subsequently).

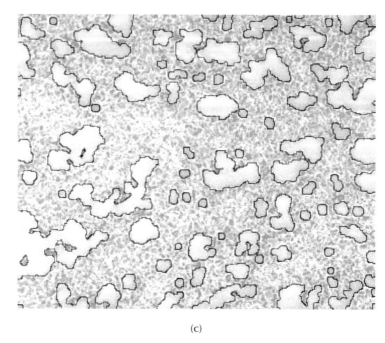

(c)

FIGURE 3.29 (continued)

Since fractal geometry has been mentioned in this context, it is an appropriate place to insert a brief digression on the measurement of fractal dimension for surfaces. These frequently arise in food products, especially ones that consist of an aggregation of similar particles or components. For example, the images in Figure 3.32 show the surface of particulate whey protein isolate gels made with different pH values. Visually, they have network structures of different complexity. This can be measured by determining the surface fractal dimension (we will see in later chapters other ways to measure the fractal dimension).

The most efficient way to determine a fractal dimension for a surface is to calculate the Fourier transform power spectrum, and to plot (on the usual log-log axes) the amplitude as a function of frequency. As shown in Figure 3.33, the plot is a straight line for a fractal, and the dimension is proportional to the slope. If the surface is not isotropic, the dimension or the intercept of the plot will vary with direction and can be used for measurement as well.

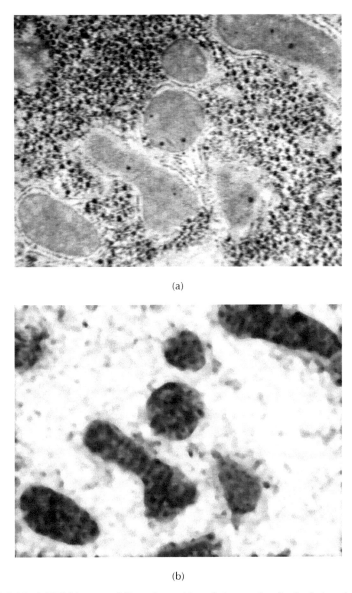

(a)

(b)

FIGURE 3.30 A TEM image of liver tissue (a) and the result of calculating the entropy within a neighborhood with radius of 4 pixels and representing it with a grey scale value (b).

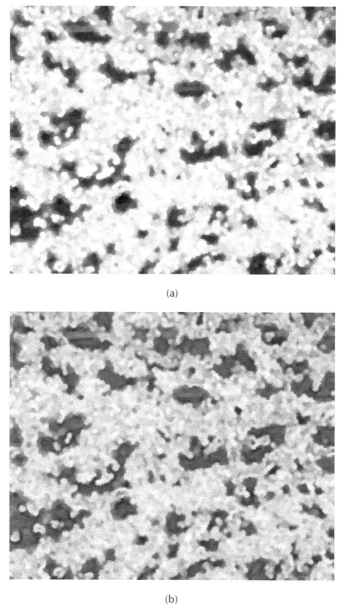

(a)

(b)

FIGURE 3.31 The slope (a) and intercept (b) from calculation of the fractal texture in the image in Figure 3.29(a).

(a)

(b)

FIGURE 3.32 Surface images of particulate whey protein isolate gel networks showing a change in fractal dimension vs. pH. (a) pH 5.4, fractal dimension 2.17; (b) pH 5.63, fractal dimension 2.45. (Courtesy of Allen Foegeding, North Carolina State University, Department of Food Science. Magnification bars are 0.5 μm)

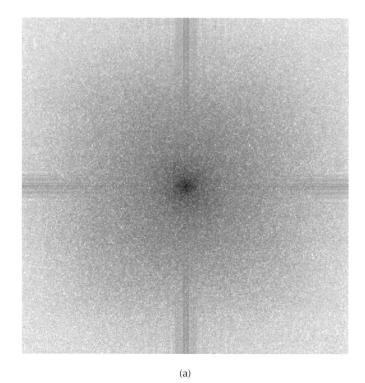

(a)

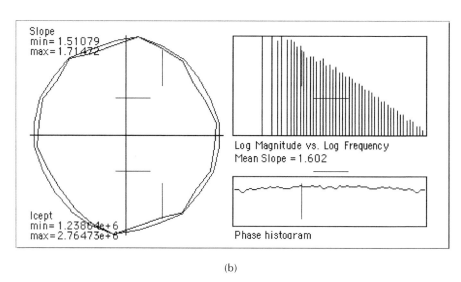

(b)

FIGURE 3.33 Analysis of the fractal gel structure shown in Figure 3.32a: (a) the Fourier transform power spectrum; (b) its analysis, showing the phase values which must be randomized and a plot of the log magnitude vs. log frequency which must be linear. The slope of the plot is proportional to fractal dimension; the slope and intercept of the plot are uniform in all directions, as shown, for an isotropic surface.

DIRECTIONALITY

Many structures are composed of multiple identical components, which have the same composition, brightness, and texture, but different orientations. This is true for woven or fiber materials, muscles, and some types of packaging. Figure 3.34 shows an egg shell membrane. Visually, the fibers seem to be oriented in all directions and there is no obvious predominant direction. The Sobel vector of the brightness gradient can be used to convert the local orientation to a grey scale value at each pixel, which in turn can be used to measure direction information. Previously, the magnitude of the vector was used to delineate boundaries. Now, the angle as shown in Figure 3.23 is measured and values from 0 to 360 degrees are assigned grey scale values of 0 to 255. A histogram of the processed image shows that the fibers are not isotropic, but that more of them are oriented in a near-horizontal direction in this region.

This method can also be used with features that are not straight. Figure 3.35 shows an example of collagen fibers, which seem not to be random, but to have a predominant orientation in the vertical direction. The direction of the brightness gradient calculated from the Sobel operator measures orientation at each pixel location, so for a curved feature each portion of the fiber is measured and the histogram reflects the amount of fiber length that is oriented in each direction.

Note that the brightness gradient vector points at right angles to the orientation of the fiber, but half of the pixels have vectors that point toward one side of the fiber and half toward the other. Consequently the resulting histogram plots show two peaks, 128 grey scale values or 180 degrees apart. It is possible to convert the grey scale values in the image to remove the 180 degree duplication, by assigning a color look up table (CLUT) that varies from black to white over the first 128 values, and then repeats that for the second 128 values. This simplifies some thresholding procedures as will be shown below.

The images in the preceding examples are entirely filled with fibers. This is not always the case. Figure 3.36 shows discrete fibers, and again the question is whether they are isotropically oriented or have a preferential directionality. The measurement is performed in the same way, but the image processing requires an extra step. A duplicate of the image is thresholded to select only the fibers and not the background. This image is then used as a mask to select only the pixels within the fibers for measurement. The random orientation values that are present in the non-fiber area represent gradient vectors whose magnitude is very small, but which still show up in the orientation image and must be excluded from the measurement procedure. The thresholded binary image of the fibers is combined with the direction values keeping whichever pixel is brighter, which replaces the black pixels of the fibers with the direction information.

The 0 to 360 degree range of the orientation values is often represented in color images by the 0 to 360 degree values for hue. Like most false-color or pseudocolor displays, this makes a pretty image (Figure 3.37) and helps to communicate one piece of information to the viewer (the orientation of the fiber), but adds no new information to the measurement process.

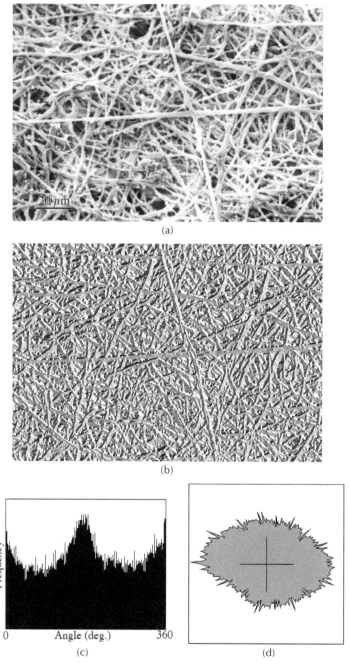

FIGURE 3.34 Measurement of orientation: (a) SEM image of egg shell membrane (courtesy of JoAnna Tharrington, North Carolina State University, Food Science Department); (b) application of the Sobel direction operator; (c) histogram of image (b), showing the relative frequency of orientations from 0 to 360 degrees; (d) data from (c) replotted as a rose of orientations (radius represents frequency).

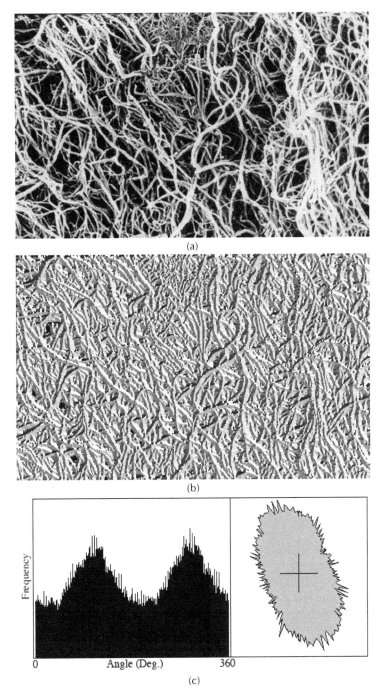

FIGURE 3.35 Measuring orientation: (a) image of collagen fibers; (b) application of Sobel orientation filter; (c) histograms of the orientation data (left, conventional 0.255 grey scale range corresponds to 0. 360 degrees; right, same data plotted as a rose plot) showing preferred orientation of the fibers.

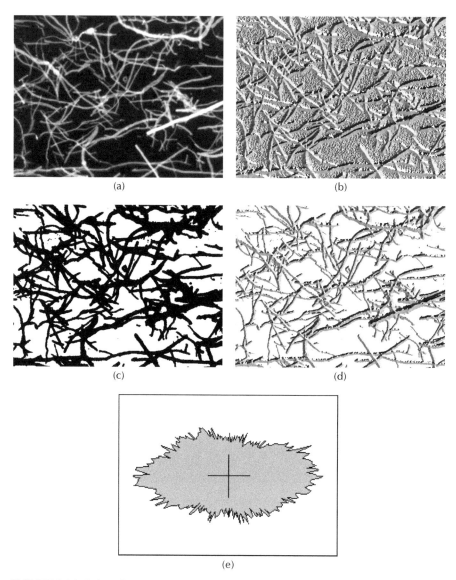

(a)

(b)

(c)

(d)

(e)

FIGURE 3.36 Orientation measurement for separated cellulose fibers: (a) original; (b) orientation value assigned to pixels; (c) mask produced by thresholding the fiber image; (d) elimination of all non-fiber pixels (note that the presence of two orientation values, different by 128 grey scale values or 180 degrees, on opposite sides of each fiber is particularly evident in this image); (e) rose plot of histogram showing preferentially oriented fibers.

Thresholding of regions in an image can be illustrated by using one of the Brodatz textures from Figure 3.28. The herringbone fabric pattern shown in Figure 3.38 is constructed with fibers running vertically and horizontally, but the visual impression is one of diagonal stripes. Applying the Sobel orientation filter assigns grey scale values to each pixel, which in this example has been reduced to a 0 to

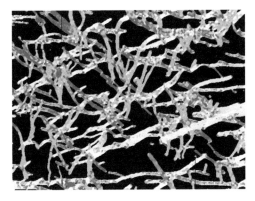

FIGURE 3.37 Assigning hue colors to the fiber angles (see color insert following page 150).

FIGURE 3.38 Example of directional thresholding: (a) original Brodatz herringbone pattern; (b) Sobel orientation; (c) noise reduction; (d) overlay of final thresholded result on original image.

180 degree range instead of the usual 0 to 360 degrees. Applying a median filter reduces the noise due to local random variations, and produces an image with two predominant brightness values, which correspond to the areas containing the two diagonal patterns. When automatically converted to a binary image as described in the section on thresholding, these can be measured to determine the dimensions, etc., of the regions.

FINDING FEATURES IN IMAGES

The herringbone textile pattern in the preceding figure is a good example of an image containing many repetitions of the same structure. While each individual chevron is slightly different from all of the others, due to the individualities of the fibers and the random noise in the image, the underlying structure is the same. By averaging together all of the repetitions, a much better image can be obtained of that structure. This is most easily accomplished by using the Fourier transform. The spikes of high amplitude in the power spectrum indicate the predominant frequencies and orientations of the terms that combine to produce the image of the repetitive structure. In the examples of removal of periodic noise in the preceding chapter, finding these spikes and eliminating those terms was used as a way to remove the periodic noise and keep the rest of the image. Now we will do the opposite: keep the periodic (structural) signal and remove the random superimposed variability and noise.

The power spectrum can be processed like any other image to locate the spikes and produce a filter that will keep them and remove other frequencies. In many cases the top hat filter is a good tool for locating the spikes, or points of high amplitude. In this particular example (Figure 3.39) it is more efficient to remove the spikes by using a rank filter to perform a grey level erosion (replace every pixel with its brightest neighbor) in order to generate a background which is then subtracted from the original. The resulting filter or mask is shown in the figure. Removal of all the frequencies and orientations that are white in the filter, and keeping those where the filter is black, allows performing the inverse Fourier transform to obtain an image in which all of the repetitions of the basic chevron structure have been averaged together, to produce a low noise average.

The Fourier power spectrum also provides a useful summary of all the periodic information in the original image that can be used for measurement. The z-bands in the muscle tissue shown in Figure 3.40 are regularly spaced but awkward to measure. In order to obtain good precision, many separate locations would have to be measured and the results averaged. The Fourier transform provides averaging for the entire image automatically. The distance from the center to the first spike in the power spectrum gives the frequency (and the orientation) of the fundamental periodicity that is present. The spacing can then be calculated as the width of the original image divided by that radial distance.

Many images contain multiple repetitions of the same structure, but they are not regularly spaced and, hence, are not represented by simple spikes in the Fourier transform power spectrum. Nevertheless, it may be possible to efficiently find them by using the frequency-space representation of the image. However, an initial understanding of the method is probably easiest using the familiar pixel version of the

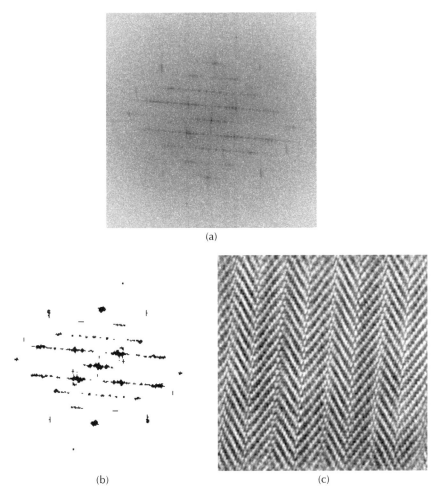

(a)

(b)   (c)

FIGURE 3.39 Example of averaging in Fourier space: (a) Fourier transform power spectrum of the original image from Figure 3.38; (b) mask created to keep just the spikes (high amplitude terms); (c) inverse transform of just the high amplitude terms producing an averaged image.

image. Imagine isolating a single representative example of the structure of interest, placing it on a transparent overlay, and sliding it to all possible positions on the original image looking for a match. As a measure of how well the target image is matched by the underlying data, the individual pixel values in the neighborhood are rotated 180 degrees, multiplied together, and summed. This cross-correlation value rises to a maximum when the target is located. It turns out that the arithmetic procedure, which must be applied with the target centered at every pixel in the original image, can be carried out very quickly and straightforwardly using the Fourier transform.

For those interested in the underlying math, the Fourier transforms of the target and the original image are both computed. These are then multiplied together, but the values are complex (real and imaginary) and for cross-correlation the phase angle

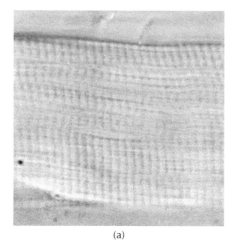

(a)

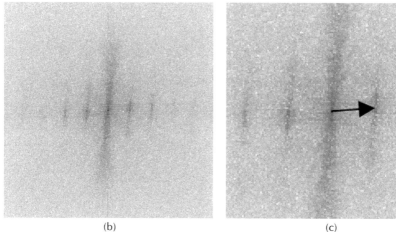

(b)   (c)

FIGURE 3.40 Measuring regular spacings: (a) original image of z-bands in muscle tissue; (b) Fourier power spectrum; (c) enlargement of the central portion of (b), showing the radial distance to the first spike. The average spacing in the original image is equal to the width of that image (256 pixels) divided by the radial distance (26.5 pixels) = 9.6 pixels. Of course, this must be used with the original image magnification to calculate the actual spacing.

of the target values are rotated by 180 degrees. Instead of requiring many multiplications for each pixel location, there is just one multiplication for each pixel. Then an inverse Fourier transform is applied to the resulting data, producing an image in which dark spots correspond to the locations where the target was matched (and the darkness of the spot is a measure of how well it was matched).

In the example in Figure 3.41, blisters on packaging are difficult to see visually (or count automatically) because they do not have simple contrast (they are bright on one side, dark on the other), and because the image is rather noisy. Selecting one of the blisters as a representative target and performing cross-correlation with the entire image marks all of the blisters for easy recognition and counting. Notice that

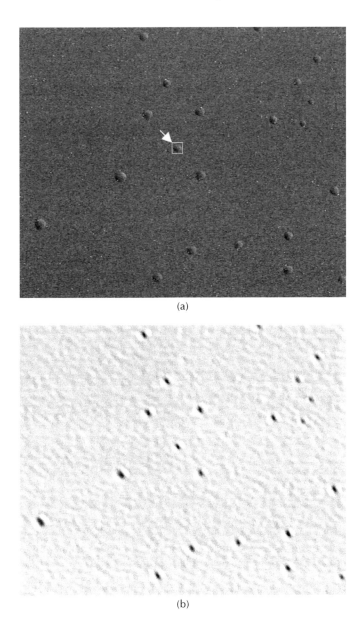

(a)

(b)

FIGURE 3.41 Surface image of packaging showing blisters: (a) the one in the center of the field (marked with a white arrow) was selected as the target. The cross-correlation result (b) marks the location of all of the blisters present for easy counting.

the sizes of the blisters vary by about a factor of 2, and so the target does not perfectly match all of the occurrences, but the marks are still dark enough to locate them all.

Cross-correlation can also see through clutter, noise, and the presence of other shapes to locate features. In the example in Figure 3.42, latex spheres have been

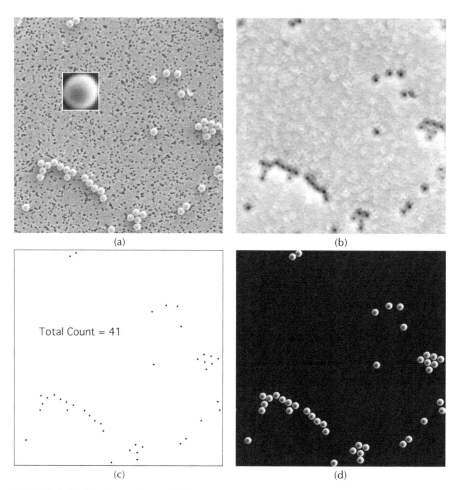

(a)

(b)

Total Count = 41

(c)

(d)

FIGURE 3.42 Nuclepore filter with latex spheres: (a) original SEM image with superimposed, enlarged target image as discussed in text; (b) cross-correlation result; (c) application of top-hat filter (inner radius = 4 pixels, outer radius = 6 pixels, crown height = 12 grey levels) and the resulting count of the number of particles present; d) convolution with the target image shows just the particles without the background or other features present in the original image.

collected from a fluid onto Nuclepore filters. The holes in the filter and the presence of other dirt on the filter make it difficult to count the particles, as does the varying contrast (isolated spheres have greater contrast than ones in groups). Because of the noise in the image, and the desire to apply the processing automatically to literally thousands of images, a target image was constructed by averaging together (manually) the images of about a dozen individual spheres from multiple original images. An enlarged copy of the target is shown in Figure 3.42(a).

Cross-correlation with this target marks all of the spheres. To simplify counting, a top-hat filter was applied to the cross-correlation result to locate all of the significant marks. This produced uniform black marks for counting. This image can then

(a)

(b)

FIGURE 3.43 Image of the surface of a particulate whey protein isolate gel network (a) (courtesy of Allen Foegeding, North Carolina State University, Department of Food Science). The autocorrelation result (b) provides a measure of the average size and shape of the structuring element (shown by thresholding a contour line on the dark spot).

be used to reconstruct a view of just the particles, without the background or dirt in the original image. This is done by convolution with the target image. (Convolution is also carried out in Fourier space, and is exactly like cross-correlation except that the phase angle of the complex values is not altered before multiplication.)

Correlation also offers a method by which the size and shape of the structural unit that is repeated in the image can be determined automatically, without selecting a target manually. In this case, the image is cross-correlated with itself (auto-correlation). As shown in the example of Figure 3.43, the result provides a direct measure of the size and shape of whatever structuring element is repeated throughout the image. In the example, a thresholded contour line has been added to show the outline of the representative feature.

To understand how this procedure works, it is again instructive to visualize an equivalent process performed with the original image made up of pixels. A transparent overlay with a copy of the image is placed on top of the original and is gradually slid away from perfect alignment. How far can it be moved in each direction before a feature in the overlay no longer covers itself in the original? The cross-correlation mathematics calculates the degree of match and thus measures the distance as a function of direction, to produce an averaged feature representation, even though many of the features in the original image are partially covered because of the three-dimensional nature of the surface. In the example shown, the size of the structuring element in the protein isolate gel network was measured in this way and shown to vary by about a factor of 2 as the $CaCl_2$ concentration was varied from 30 to 50 mM.

Cross-correlation has many other uses in image processing. One is to match points in stereo pair images so that the lateral displacement (disparity) can be determined and used to define the elevation of points. Depending on the type of surface being imaged and the imaging method, this may be done either for every point in each image or for just a few thousand interesting points (locations where the variance is high and locally maximum), which are then used to interpolate a smooth model for the surface. Cross-correlation is also used in tracking moving features in a sequence of images, and for image alignment.

IMAGE COMBINATIONS

Several examples of the arithmetic combination of two images have already been used in this chapter and the preceding one. For example, the removal of nonuniform brightness by subtracting or dividing by a background image is a fairly common procedure. In Figure 3.36 above, a binary image of the fibers was combined with the result of the Sobel direction filter to keep the brighter pixel. This replaced the black fibers with the direction value, while keeping the white background.

A full complement of arithmetic operations (addition, subtraction, absolute difference, division, multiplication, and keeping either the brighter or darker value) is usually available, operating pixel by pixel to combine two images. Subtraction and division, in addition to their use for background leveling, are often used to compare images acquired through different colored filters or different imaging modes (e.g., in the AFM or SEM). Addition is primarily used as a way to compare the results of thresholding to the original image, as shown in Figure 3.38d above. It may also be used to combine multiple wavelength images, for example a representation of the visual monochrome brightness in a color image is often modeled as approximately 0.65 * Green + 0.25 * Red + 0.10 * Blue. Multiplication is infrequently employed in image processing but is used in computer graphics to superimpose a texture onto surfaces. Keeping the brighter or darker pixel values will be used extensively in the next chapter to combine thresholded binary images with grey scale values. We will also encounter there the various Boolean combinations (AND, OR, Ex-OR) that can be used with two binary images.

All of these combinations require, of course, that the images be aligned with each other. If they have originated from the same initial image, for instance by different processing operations, then proper alignment is assured. If several original images have been acquired, for example by recording different wavelengths, some adjustment for misalignment may be required. In some cases, including medical imaging with different modalities such as PET, CT, and MRI, or remote sensing images recorded by different satellites, the image magnifications, orientations, and even points of view may be different.

For simple translation alignment of multiple images, the cross-correlation procedure shown above works automatically and quickly to provide the desired result. If two similar images of the same subject are cross-correlated, the result is a peak whose maximum is displaced from the center by exactly the amount of misalignment. The center of the peak can be located by interpolation to a fraction of a pixel, and the image shifted by that amount to produce proper alignment. Figure 3.44 shows an example. This technique is also useful for aligning stereo pair images for viewing.

This type of shift alignment is also useful when dealing with video images of moving subjects. The interlace scan used in standard video records each of the thirty frames per second in two halves, or fields. These capture the even and odd scan lines, respectively. If the scene is not stationary (either because the subject or the camera is moving) this produces an offset as shown in Figure 3.45. Performing cross-correlation between the two fields of the image allows correction of the interlace offset. The assumption is that the entire scene is shifted uniformly; if different objects within the field have different motions the result is an average.

For situations in which shift, rotation, and perhaps perspective correction are all needed to bring two images into proper registration, the basic procedure is the same as for the correction of image distortion, shown in the preceding chapter. Generally this requires the location of some fiducial marks in the images. In a few cases this can be handled automatically. This may be possible in medical imaging by using a few distinctive structures known to be present. A more common technique places markers in the image, for instance by embedding wires or fibers, or drilling holes in a block of embedded material before cutting sections for microscopy. This technique is primarily used for alignment of serial section images for combination or comparison.

The most common procedure is to rely on human knowledge and recognition, and allow the user to mark registration points on the images. A minimum of one point is needed for translation, two for translation and rotation, three for translation, rotation, and scaling, and four to also include perspective correction. Some programs

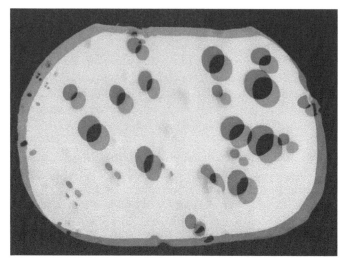

(a)

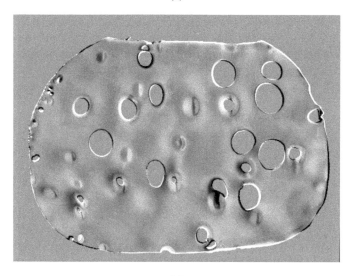

(b)

FIGURE 3.44 Cross-correlation shift: (a) overlay of scans of two sides of a slice of swiss cheese (one image flipped side-to-side); (b) after cross-correlation alignment, subtracting one from the other reveals the holes that do not extend all the way through the slice, which are counted for use in the disector, a stereological routine that measures the number of holes per unit volume.

allow more points to be used either to determine the best fit values of the needed parameters, or to perform local image morphing to match one image to another. Morphing is not usually used as part of an image analysis procedure, but it may be useful in medical imaging to align the image of a subject to a standard template for visual comparison.

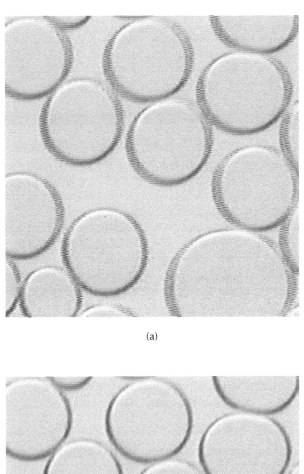

(a)

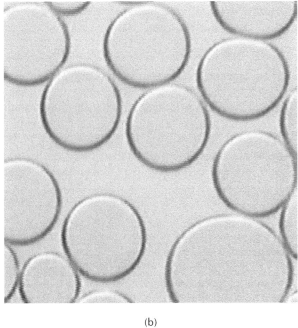

(b)

FIGURE 3.45 Correcting interlace shift: (a) original video image of bubbles in moving fluid; (b) after cross-correlation alignment of the even- and odd-numbered lines.

In the illustration shown in Figure 3.46, the most common case of alignment requiring translation, rotation and scaling is shown. Three points are marked on each image, which define the required vectors. The result is a set of equations that give the X, Y coordinates in the original image for each point X, Y in the transformed (aligned) image. These are typically of the form

$$X = a + b\,X + c\,Y$$
$$Y = d + e\,X + f\,Y$$

(3.2)

and produce values for X and Y that are not integers. Hence it is necessary to interpolate in the original image to determine the pixel brightness value to be assigned to the new image. In fact, interpolation is commonly required for all of the alignment or perspective correction procedures.

The simplest interpolation method is to use the nearest neighbor value, simply truncating or rounding off the X, Y values to integers and using the corresponding original pixel value. This preserves brightness data best, but causes aliasing of edges as shown in Figure 3.47. Bilinear interpolation between the four nearest pixels, or bicubic interpolation fitting functions to the nearest 16 pixels, produce visually superior results. More complex interpolation methods using sinc, spline, or other functions are sometimes used in computer graphics but offer little advantage in routine image processing applications.

Once images have been aligned, the various arithmetic operations can be applied to combine them. In addition to these steps, the use of color channels (either RGB or HSI) may be useful to reveal complex relationships. In the example shown in Figure 3.48, both of these tools are used. The Fura stain reveals calcium activity. Ratioing the image acquired with an excitation below the absorption edge of the stain to that with an excitation above the edge measures the amount of calcium activity. But the total amount of stain also has a distribution in the sample, which can be measured with the sum of the two images. One way to show these results together is to place the ratio image into the hue channel and the sum image into the intensity channel of a color image (in the example, saturation is set to maximum).

As another example of image ratios and combining images for display, the ratio of two fluorescence images acquired at different times can be used to calculate lifetimes. The ratio image is commonly displayed in the hue channel while the intensity channel shows the original intensity distribution.

Tracking the motion of objects also uses image combinations. In the example of Figure 3.49 a series of images taken over a period of time are combined by keeping the darkest pixel at each location (because the features are darker than the background). The combined image can be thresholded and skeletonized as described in the next chapter to mark the path of each feature for measurement.

The most common purpose for image combinations is the detection of image differences, usually by subtraction or ratioing. The differences may reflect changes that have occurred over a period of time, or as imaged with different wavelengths. Either serial physical or optical sections produce images from different depths in the sample which can be compared for stereological calculations. Many of these procedures are shown in this and other chapters in the context of the analysis.

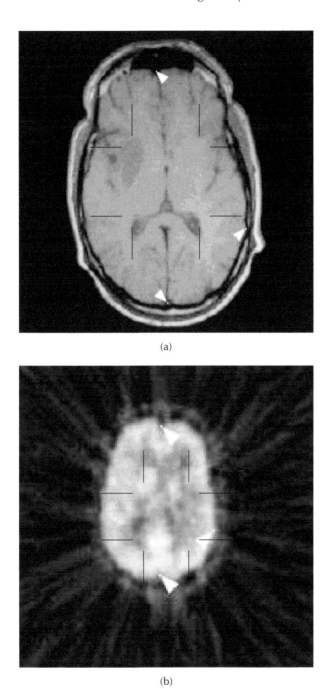

(a)

(b)

FIGURE 3.46 Three-point registration: (a) MRI; (b) PET; (c) CT scans of a human head. The original three images are different in size and pixel resolution as well as orientation, and three registration marks (arrows) were placed by hand based on visual judgment; (d) after alignment, the images were combined as RGB color channels for viewing (see color insert following page 150).

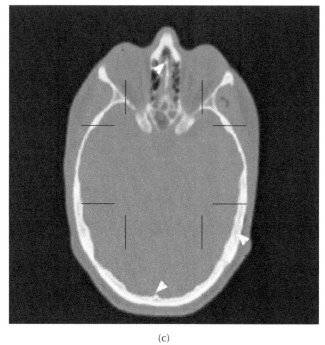

(c)

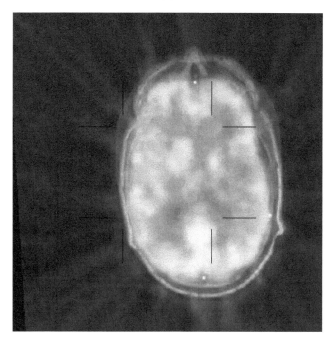

(d)

FIGURE 3.46 (continued)

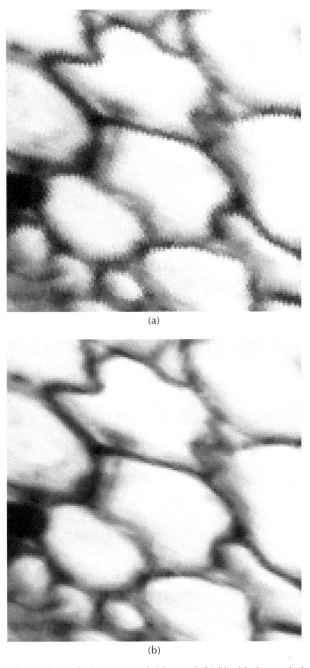

(a)

(b)

FIGURE 3.47 Comparison of (a) nearest neighbor and (b) bicubic interpolation after image scaling and rotation. Original image is a fragment of Color Figure 3.57; see color insert following page 150.

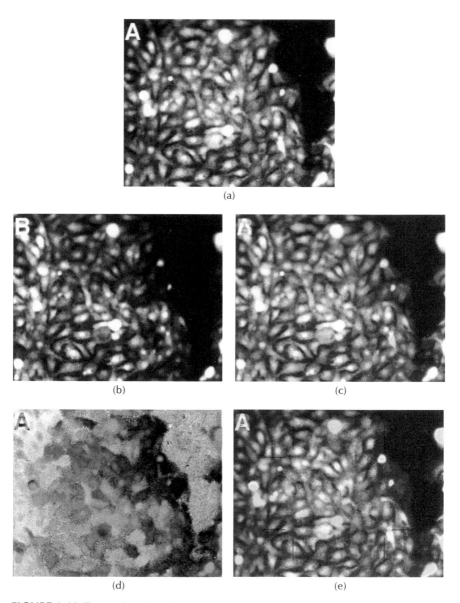

(a)

(b)   (c)

(d)   (e)

FIGURE 3.48 Fura stain to localize calcium activity: (a) excited at 340 nm; (b) excited at 385 nm; (c) sum; (d) ratio; (e) color composite as described in the text (see color insert following page 150). (Courtesy of P. J. Sammak, University of Minnesota)

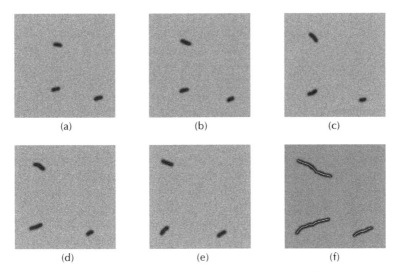

(a)   (b)   (c)

(d)   (e)   (f)

FIGURE 3.49 Feature tracking by image combination: (a to e) individual images; (f) composite with superimposed skeletons showing the paths of motion.

THRESHOLDING

The image processing described in this chapter and the preceding one may be performed for several different reasons. Improving the visual appearances of images can assist the researcher in detecting and observing important details. Processing is also used in preparing images for publication and to more effectively communicate the important details to others who are less familiar with the subject. Processing can be used to produce some measurement information directly, often by analysis of the histogram. The other reason for processing is to prepare images for thresholding (and subsequent quantitative measurement) of the features or structures present, by producing an image in which brightness or color values are uniquely assigned to the different structures. In the next chapter, processing and editing of the binary images produced by thresholding is described, which is an indication that thresholding is often an imperfect and not a final step toward delineation of the structures to be measured.

Thresholding in its simplest form is the process of selecting the pixels that make up the structure of interest, on the basis that they have a unique brightness or color range. These foreground pixels are then colored black and the remainder, the background, are colored white, producing a binary image. A word of caution: some programs use the reverse convention in which the foreground is white and the background black. Generally this indicates that they originated when computers displayed white (or green, or amber, depending on the phosphor color) letters on a dark cathode ray tube, and before the now-common white screens and black letters came into widespread use. It makes no difference which convention is used (as long as you know which is which). In any case, binary images are those in which there are two kinds of pixels, which make up the features and the background.

All image analysis systems incorporate some form of thresholding (also sometimes called segmentation, although that term will be reserved here for separation of touching features). The most basic form of thresholding is simply the manual setting of a brightness level to distinguish brighter from darker pixels. As shown in the example of Figure 3.50, this is usually done by referral to the image histogram. The problem is that the histogram contains very little information to help adjust the setting properly.

One of the very basic structural relationships presented in the chapter on stereology is that the area fraction of a phase or structure measures the volume fraction. In the example, the histogram represents only the pixels in the meat, excluding the black background. The broad peak represents the dark meat and the wide flat shelf represents the bright fat and bone. It is not obvious where the threshold should be positioned. Threshold settings in the range from about 105 to 130 (shown on the histogram) all produce visually plausible binary images, but the area fraction measured for the meat ranges from 71 to 80%. Using an automatic setting of 110 based on a statistical test (discussed below) produces a measurement of 73% meat, which corresponds closely to the value reported by physical tests.

In the more general case, there are two threshold values that can be set to select a range of brightness values. And in many real situations, as shown in Figure 3.51, it is the background that is the most uniform and can be selected while the features include both brighter and darker values. The histogram shows a peak for the grey

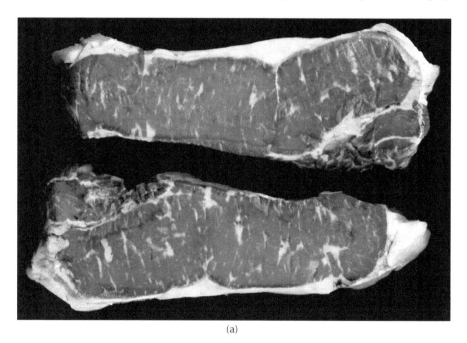

(a)

FIGURE 3.50 Example of threshold setting: (a) scanned image; (b) histogram showing a range of settings from 105 to 130 (on a brightness scale of 0 to 255); (c) thresholded binary image of fat produced by a setting of 110.

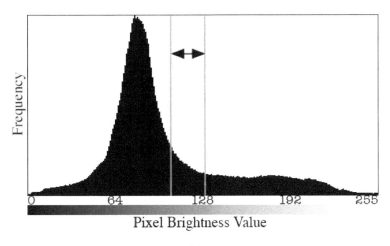

Pixel Brightness Value

(b)

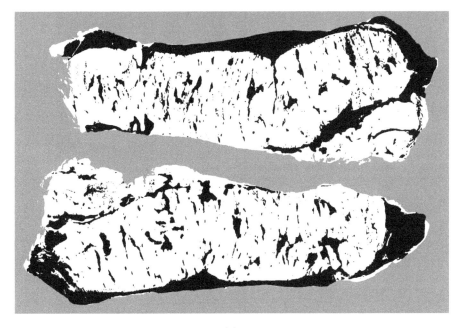

(c)

FIGURE 3.50 (continued)

background around the emulsified fat droplets in this image of milk, but as before there are no obvious places to set the threshold values. Also note that the resulting thresholded image has many of the droplets touching each other. Segmentation of touching features after thresholding is discussed in the next chapter.

The examples of thresholding in textbooks often show an idealized histogram of the type in Figure 3.52, with well separated peaks that invite the setting of

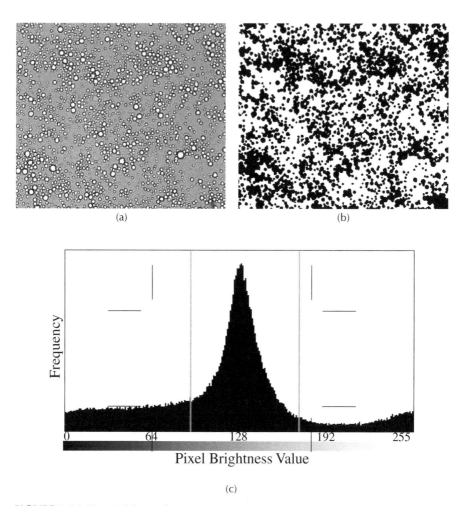

(a)   (b)

(c)

FIGURE 3.51 Thresholding using two levels and the uniformity of the background: (a) original (milk, showing emulsified fat droplets; courtesy of Ken Baker, Ken Baker Associates); (b) thresholded binary image inverted to show the droplets; (c) histogram with the upper and lower threshold settings spanning the grey values for the background.

thresholds at the minima in the valleys between the peaks, or sometimes halfway between the peaks. Sometimes these methods work for materials samples, or for quality control applications where the goal is reproducibility rather than accuracy, but they are fundamentally flawed for several reasons. First, few images of food or other organic material produce such simple multimodal histograms. Usually the appearance is more like the examples just shown, with poorly defined peaks (or no peaks at all).

Second, the midpoint between peaks has no particular meaning, because as noted in the preceding chapter the acquisition device may be linear, or logarithmic, or have some other output characteristic. And thirdly, the minimum point is not stable but moves around. Consider the case of a two phase structure with well-defined peaks,

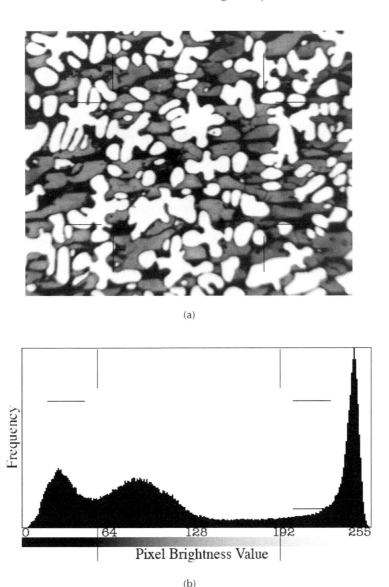

(a)

(b)

FIGURE 3.52 Example of a micrograph of a metal alloy with three well-defined and homogeneous phases, and its histogram showing the three corresponding, well-separated peaks. Most organic materials do not have such simple brightness distributions.

as indicated in Figure 3.53. Without changing the illumination or camera settings, simply scan the field of view to an region of the sample where the area fractions are different, and as shown in the example the peak's heights will change, and with them the location of the minimum. There is no reason for the correct threshold to vary in such a case, so clearly using the minimum is not a correct strategy.

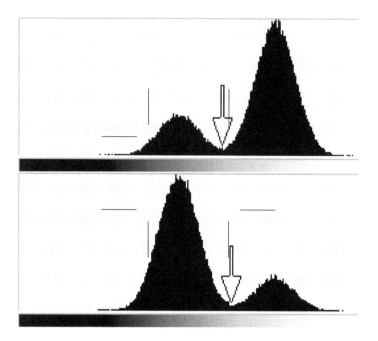

FIGURE 3.53 Changing the relative peak heights of two structures by varying their area fraction as described in the text. The peak positions do not change, and the correct threshold should not change, but the minimum between them shifts by more than 10 brightness levels.

AUTOMATIC THRESHOLD SETTINGS
USING THE HISTOGRAM

Manual setting of threshold levels is usually accomplished by adjusting sliders on the histogram while visually observing the image, where some sort of preview shows which pixels have been selected. Moving the sliders until the result looks right is difficult, particularly when the image is large and must be scrolled to see various regions. More important, what looks right to one person may not to another, or even to the same person on a different day or if the image is rotated. Thresholding images is the step where most image measurement errors arise, because of inconsistent human judgment. Consequently, there has been a continuing effort to find algorithms that can decide where the threshold values should be set.

All of these techniques depend upon knowing something about the image, how the sample was prepared and the image acquired, and what structures or features should be present. This information is presumably what the user relies upon in deciding that the manual settings look right, but they must be made explicit in order to determine the criteria which an automatic procedure can apply.

The greatest number of automatic thresholding techniques, and certainly the most widely used methods, apply to the specific case of printed text on paper (usually thresholding in this circumstance is a precursor to character recognition and conversion of the image to a text file). The independent knowledge about the image in this case is that the image consists ideally of just two kinds of pixels — ones that are generally bright and correspond to the paper, and ones that are darker and correspond to the ink. With that criterion, a variety of statistical classification methods have been developed.

Figure 3.54 shows two of the more generally successful and widely used approaches. Both assume that the histogram actually consists of two distributions,

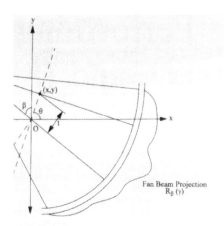

Differentiating Eqs. ...

$$\frac{\delta t}{\delta \gamma} = D(\beta) \cos \gamma, \qquad \frac{\delta \theta}{\delta \gamma} = 1,$$

$$\frac{\delta t}{\delta \beta} = D'(\beta) \sin \gamma, \qquad \frac{\delta \theta}{\delta \beta} = 1.$$

Using the Jacobian and the th

$$dt \, d\theta = [D(\beta) \cos \gamma - D'(\beta) \sin \cdot$$

Performing a change of variables (7), and (8), we have

$$f(r, \phi) = \frac{1}{2} \int_0^{2\pi} \int_{-\gamma_m}^{\gamma_m} P_{\beta+\gamma} [D(\beta)$$

$$\times [r \cos(\beta + \gamma - \phi) - D$$

$$\times [D(\beta) \cos \gamma - D'(\beta)$$

fan-beam geometry. Fan-beam projections are rotation angle β and sampling angle α.

are generally satisfied. In particular, the holds when the source-to-origin distance

(a)

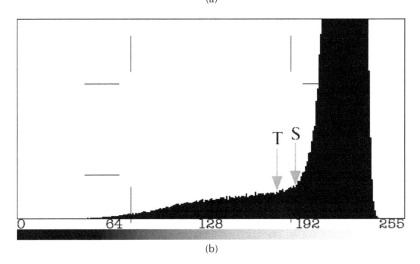

(b)

FIGURE 3.54 Example of print on paper: (a) original; (b) histogram with threshold values selected by the Trussell (T) and Shannon (S) algorithms; (c) Trussell result (detail); (d) Shannon result (detail).

$$dt\ d\theta = [D(\beta \quad dt\ d\theta = [D(\beta$$

Performing a Performing a
(7), and (8), (7), and (8),

(c)   (d)

FIGURE 3.54 (continued)

which may be somewhat overlapped (some ink pixels may be brighter than some paper pixels) but which can be best separated by setting a single threshold value. The definition of best is a function of what statistical test is applied. For example, the Trussell method uses the statistician's t-test to compare the two populations divided by every possible threshold setting to calculate the t-statistic. This is a function of the mean values, standard deviations, and number of pixels in each segment of the histogram. When the t-statistic is maximum the probability that the two populations are different is greatest, so that is the threshold value used.

The Trussell method works quite well for most text-reading applications, and in fact it is generally applied to many situations where it is known beforehand (or assumed) that there are just two populations of pixels present. That is often the case for stained tissue, for example (stained vs. not stained). It may also apply to porous material (solid or void), and a variety of meat and vegetable products. It is surprising that it works so well because one of the underlying assumptions in the t-test is that the populations of pixels have brightness values that have a normal (or Gaussian) distribution, so that the mean and standard deviation fully characterize the data. Few real images (even ones of print on paper) present histograms consisting of Gaussian peaks. Note in the figure that the histogram has one more-or-less symmetrical peak for the white paper pixels, whereas the darker ink pixels do not produce a peak at all, but rather a broad sloping shelf with no obvious place to position a threshold.

There are a variety of nonparametric statistical tests that can be applied to the two-population model that do not make an assumption of normality, and most of them have been used to program other threshold-selection algorithms. The Shannon method, for example, calculates the entropy for the two populations that are treated as fuzzy sets to determine the threshold setting that minimizes the uncertainty of the setting. This method selects values that are typically slightly different from the Trussell method, but also generally satisfactory. All of the examples of thresholding based on the histogram that follow in this text use either the Trussell or Shannon method. Many of the other methods (whose mathematical algorithms are summarized

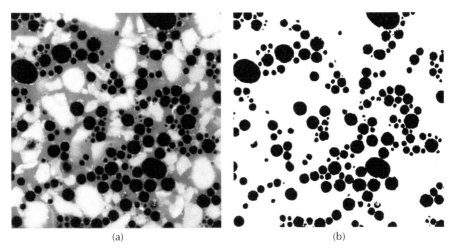

(a)   (b)

FIGURE 3.55 Stained mayonnaise: (a) original (CSLM image courtesy of Anke Janssen, ATO B.V. Food Structure and Technology); (b) after automatic thresholding.

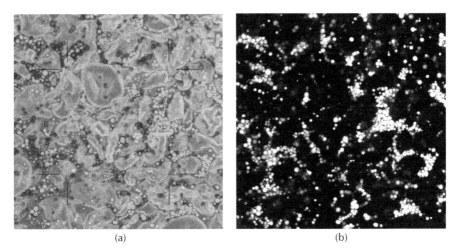

(a)   (b)

FIGURE 3.56 Stained custard: (a) original (see color insert following page 150; CSLM image courtesy of Anke Janssen, ATO B.V. Food Structure and Technology); (b) red channel selected for thresholding.

nicely in J. R. Parker, *Algorithms for Image Processing and Computer Vision*, John Wiley & Sons, New York, 1997) also work well for the print on paper application, but sometimes produce quite bizarre results on other images.

Chemical staining of tissue or food samples often produces images that can be approximated as having two populations of pixels (with and without the stain). Consequently, the automatic methods just described are often useful for thresholding these as well. In Figure 3.55 the fat droplets in rhodamine-stained mayonnaise appear as dark circles. Automatic thresholding distinguishes them from their surroundings.

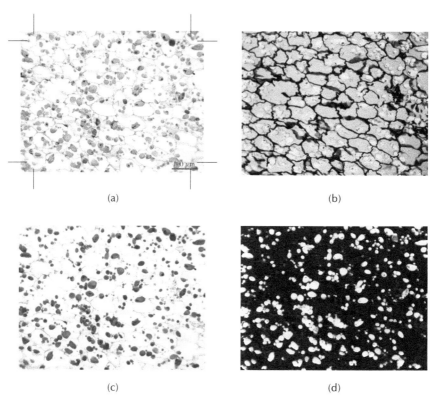

FIGURE 3.57 Stained potato: (a) original (see color insert following page 150); (b) red channel (showing cell walls); (c) green channel (showing starch granules); (d) hue channel (better contrast for starch granules).

In order to measure their size distribution it is necessary to separate the touching features. These topics are covered in subsequent chapters.

In many cases, color images produced by chemical staining can be reduced to a two-population problem, so that automatic thresholding can be applied, by first separating the image into appropriate color channels. For example, the size distribution and clustering of the Nile red stained fat in the custard in Figure 3.56 can be measured after selecting the red channel for thresholding. The starch granules and cells in the light micrograph of a stained section of potato (Figure 3.57) can each be isolated in the green and red channels, respectively.

Although RGB channels can often be used for this purpose, in general it is HSI space that corresponds better to the discrimination of structures produced by chemical staining. The color of the stain is represented by the hue value, the amount of stain by the saturation, and the density of the tissue by the intensity. In Figure 3.57(d), isolating the hue channel produces a grey scale image with much cleaner depiction of the starch granules than one of the RGB color channels. This is especially true when two different color stains have been applied. In Figure 3.58 an H&E stain produces two colors (orange and cyan) that are easily distinguished and automatically thresholded in the hue channel using the two-population assumption.

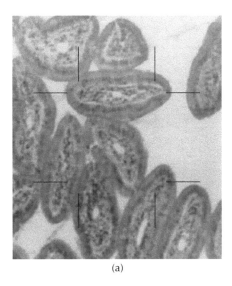

(a)

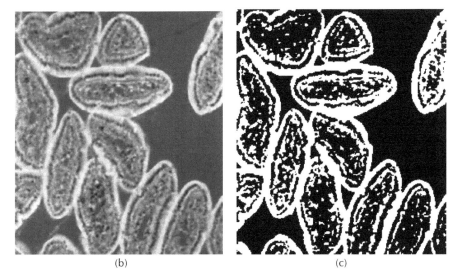

(b)   (c)

FIGURE 3.58 Stained intestine: (a) original (see color insert following page 150); (b) hue channel; (c) automatic threshold setting.

In several of the examples above, simple thresholding will not produce the final measurable result. For example, the fat droplets in mayonnaise and custard must be separated (as discussed in the next chapter). In some cases, such as the measurement of the cell wall area in the potato, the need for subsequent processing of the

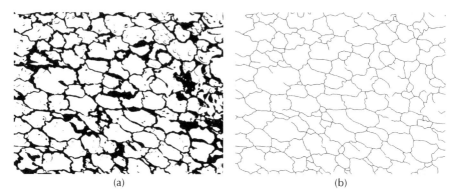

(a)   (b)

FIGURE 3.59 Thinning after thresholding: (a) thresholded red channel from Figure 3.57(b); skeletonized (and short branch segments discarded).

thresholded binary image relaxes the requirements for thresholding. In Figure 3.59 the thresholded red channel (Figure 3.57b) is skeletonized to thin lines whose total length is related stereologically to the surface area of the cells. Since the skeletonization (discussed in the next chapter) is used to thin down the cell walls, the setting of the threshold (which controls to some extent the apparent thickness of the stained tissue) can be varied over a considerable range without affecting the final result.

Of course, not all images contain just two populations of pixels. The statistical tests can be extended to deal with more groups, provided that the actual number of distinguishable structures is known. Figure 3.60 shows an example. The original image has visually distinguishable regions that have the same brightness values (both mean and distribution), but the texture has an orientation. Applying a Sobel orientation filter as described above assigns grey scale values from 0 to 255 as the direction of the brightness gradient vector varies from 0 to 360 degrees. By using a color lookup table (CLUT) as described previously, the pairs of direction values that are 180 degrees apart in Figure 3.60(b) can be mapped to the same value, and the 0- to 360-degree range of directions converted to 0 to 180 degrees, as shown in Figure 3.60(c). This image has a histogram with three well-defined peaks. Generalization of the student's t-test described above to more than two populations results in the familiar analysis of variance (ANOVA) statistical test. Applying that with the assumption of three populations of pixels calculates threshold values that separate the regions.

This example is also a reminder that the thresholding step generally occurs after various stages of image processing. If nonuniform illumination is present, or excessive image noise, or other image acquisition defects, thresholding cannot be successfully accomplished until those problems have been corrected. If enhancement is required to convert some characteristic of the image that can be visually distinguished to a brightness difference so that thresholding is possible, that must be done first.

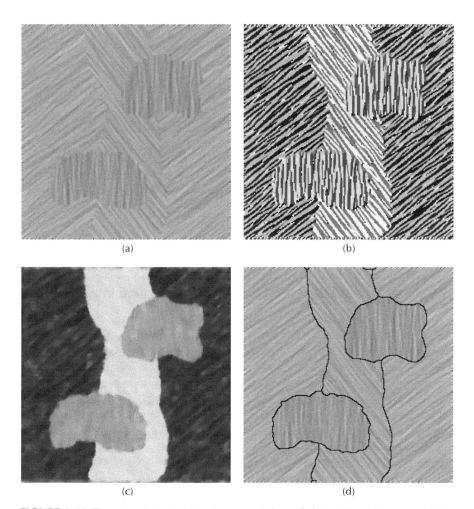

(a)   (b)

(c)   (d)

FIGURE 3.60 Example of thresholding three populations of pixels: (a) test image containing three orientations of texture; (b) application of Sobel orientation filter; (c) conversion of 0 to 360° angle values to 0 to 180° values; (d) boundary lines between regions obtained by automatic thresholding using an ANOVA test; (e) histogram showing three well-separated peaks, and the threshold settings calculated using an ANOVA statistical test used to produce the boundaries shown in image (d).

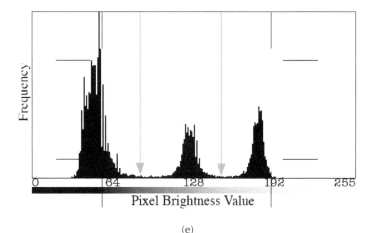

(e)

FIGURE 3.60 (continued)

AUTOMATIC THRESHOLDING USING THE IMAGE

All of the methods described above use the image histogram as a starting point, and treat the pixel values as individual and independent bits of information. In particular, they ignore the relationship between a pixel and its neighborhood. Most pixels in an image represent the same structure as their neighbors, and have similar brightness values. Only a few lie adjacent to the boundaries between one structure and another, or between feature and background. When we perform manual threshold adjustment, rather than observing the histogram, most of the attention is directed toward the appearance of the resulting image. The boundaries of the thresholded binary representation of the structure must agree with what is seen and what is known about the structures.

One of the most commonly encountered situations is images in which the feature boundaries are locally smooth. For liquids, such as the images in Figures 3.55 and 3.56, surface tension creates a force that makes the droplets more-or-less spherical and makes the surfaces locally smooth. In tissue, cell membranes do the same. Many structures that form by slow growth have locally smooth boundaries because diffusion along the boundary tends to fill in any irregularities. Wear on surfaces can also result in smooth shapes, as can local melting. When it is expected that the boundaries should be smooth, the process of adjusting a threshold setting involves watching the image and moving the slider so that the irregularities along the boundary are minimized. An algorithm that finds the threshold setting such that minor changes in threshold produce the least change in the total length of the perimeter can automate the process.

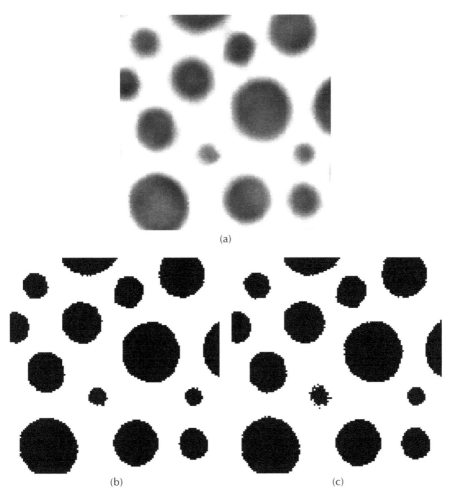

FIGURE 3.61 Smooth boundary thresholding: (a) detail of image of bubbles in a whipped food product (courtesy of Allen Foegeding, North Carolina State University, Department of Food Science); (b) binary image resulting from automatic threshold setting for smoothest perimeter; (c) binary image resulting from changing the threshold setting from 121 to 125.

Figure 3.61 shows an image of bubbles in a whipped foam food product (the full image was shown in Chapter 2, Figure 2.19). Setting an accurate threshold is difficult because of the intermediate grey values that lie around the periphery of the bubbles (due to the penetration of light into the foam so that the image represents a significant depth in the material). It is reasonable to expect the bubbles to have smooth borders, due to surface tension. Thresholding this using the smoothest perimeter criterion produces a setting of 121 on the 0 to 255 scale. Changing that setting even slightly (a value of 125 is shown in the figure) makes the borders of the bubbles significantly more irregular.

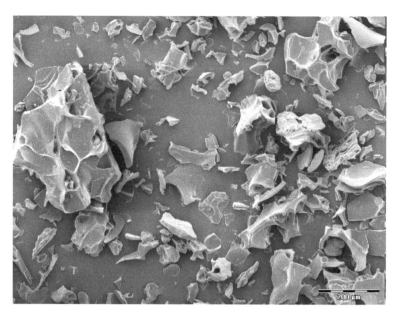

FIGURE 3.62 SEM image of pregelatinized cornstarch. Brittle fracture has produced highly variable, irregular shapes.

For images in which the boundary is sharp and has good contrast to begin with, this method usually produces results that agree closely with what a skilled and knowledgeable human operator determines. The ability to produce reasonable results as the image quality degrades is a strong plus for this method. But its use depends on knowing *a priori* that the boundaries should be smooth, and this is by no means a universal condition. Figure 3.62, for instance, shows an example of particles that have very rough boundaries. In general, objects produced by fracture and some forms of agglomeration have rough, even fractal borders. In some cases using a criterion of maximum, rather than minimum, perimeter change with threshold setting can successfully produce automatic threshold values.

Many other criteria can be used to generate automatic thresholding algorithms, but all require independent knowledge about the nature of the sample, which in most cases must come from information external to the image. It is often useful to try to make such knowledge explicit by forcing yourself to write down every detail you can think of about the specimen, its preparation, and the imaging process. If this information is sufficient to allow someone else to obtain the same images, results, and understanding that you have, then probably the list contains the information that would enable successful thresholding of the features.

Figure 3.63 shows an example in which it can be anticipated that the features (nuclei in frog red blood cells) should all be about the same size. Applying that criterion to the thresholding process produces the threshold setting shown). It has no obvious relationship to the shape of the histogram.

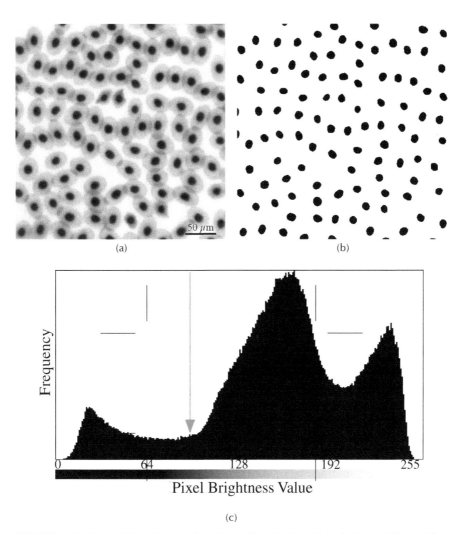

(a)   (b)

(c)

FIGURE 3.63 Threshold setting based on size uniformity: (a) original (frog red blood cells); (b) thresholded binary result; (c) histogram with threshold setting.

OTHER THRESHOLDING APPROACHES

There are many other techniques that are used, particularly in specialized applications, to separate the features of interest in an image from each other or from the background. One of these, often used in robotics and machine vision, is a split-and-merge strategy. Starting with the entire image, the histogram is examined. If statistical tests indicate that it is not composed of a single class of pixels, the image is split, usually into four parts (from which the name "quadtree" is also taken to apply to this method). The same uniformity test is repeated on each part, so that any regions that are not homogeneous are ultimately split down to the individual pixels. At the same time, after each iteration of splitting, regions are compared to their neighbors and joined (merged) if the same statistical test indicates similarity. The method is reasonably efficient, but depends critically on the ability to detect the presence of inhomogeneity within a region from the histogram. If the inhomogeneity represents only a small fraction of the area, it is unlikely to be detected.

Another approach that is complementary to traditional thresholding draws contour or iso-brightness lines on the image. Instead of using the logic that a peak in the histogram indicates the similarity of the pixels represented, this method examines the valleys in the histogram as indicators of boundaries between regions. Contour lines have several advantages for separating the structures in an image. For one thing, they are continuous lines that close on themselves (or run off the edge of the image). For another, it is not necessary for a pixel to have any particular value to locate the contour line. It follows a path between values that are greater and ones that are smaller than a given threshold value. Finally, the contour lines very often correspond to the visually detected boundaries in images, even for rather complex structures.

Figure 3.64 shows a simple contour line applied to a scanned-probe image of a coin. In this case, the brightness of each pixel represents a physical elevation. Consequently, the contour line really is a line of constant elevation and as such can mark the boundaries of the raised figure on the surface.

It is easy to extend this method to draw multiple contour lines, converting the image to a contour map in which the lines have exactly the same meaning as on a conventional topographic map. In the example of Figure 3.65, a set of ten lines equally spaced in brightness or elevation are drawn using a scanned probe image of the tip of a ball point pen. The irregularities of the lines show clearly the roughness of the ball, their spacing can be used to measure its curvature, and the elongation of the circles into ellipses measures the out-of-roundness of the ball. All of this same information is, of course, present in the original image, but it is much more visible and more easily accessed for measurement purposes in the contour map.

One of the most useful attributes of the contour method is the outlining of boundaries around features, which facilitates measurements of structure or features. In Figure 3.66 a typical image of bread shows a variety of pore sizes and shapes. Drawing a single contour line on this image produces an outline of the pores. The same logic for determining the optimum threshold setting was used as for binary thresholding (in this example the Shannon algorithm). As discussed in Chapter 1, the resulting lines provide a direct measure of the total surface area of the pores (the surface area per unit volume of the bread, which is a major factor in the mouth-feel and texture of the bread, is equal to $4/\pi$ times the total length of the contour lines divided by the area of the image).

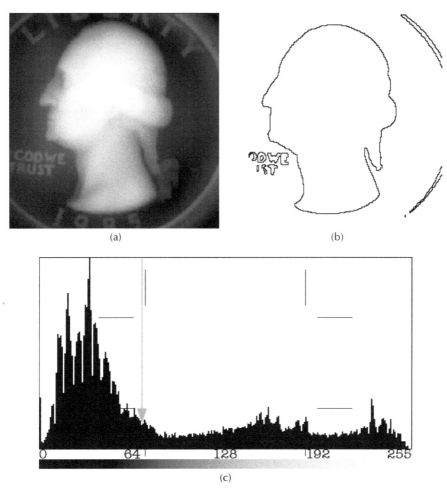

(a)

(b)

(c)

FIGURE 3.64 Drawing a contour line on an image: (a) scanned probe image of a coin; (b) contour line drawn to outline the raised head; (c) histogram of image (a) with the brightness value used for the contour line marked.

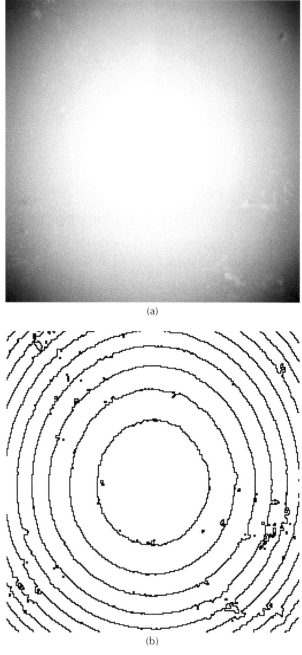

(a)

(b)

FIGURE 3.65 Drawing a contour map: (a) scanned probe image of a ball point pen; (b) contour map drawn from the image.

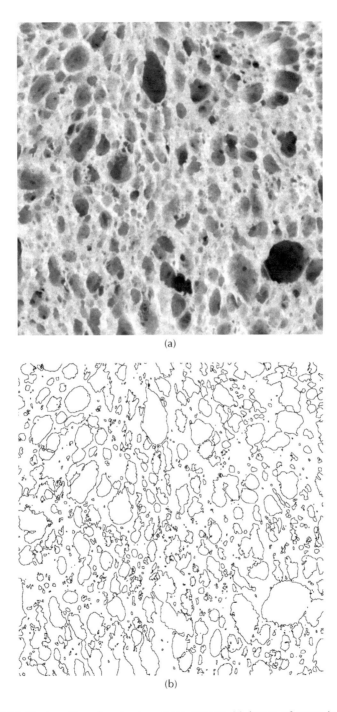

(a)

(b)

FIGURE 3.66 Contour lines for structure measurement: (a) image of pores in bread; (b) contour line drawn to mark the edges of the pores.

COLOR IMAGE THRESHOLDING

In most cases, color images are thresholded by setting threshold values on one or several of the individual color channels (RGB or HSI). In cases requiring more than one channel, the results are combined using a Boolean AND. In other words, the yellow candies in Figure 3.67(a) could be selected by separately thresholding the red channel for all of the features containing a high red intensity (the red, orange and yellow objects), and the green channel for all features containing a high green intensity (the green and yellow objects), and then combining the images to keep only those pixels that were selected in both channels. Examples of such Boolean combinations will be discussed in the next chapter.

Alternately, for a simple image such as this one, the yellow features could be thresholded by setting levels on the hue channel to distinguish the yellow objects from all of the others. That would not work, however, for the brown candies since they have the same hue as the orange ones. Combining that selection with limits on the saturation (brown is less saturated than orange) and/or intensity (brown is darker than orange) would require thresholding the H, S, and I channels separately and performing a Boolean AND to keep just the pixels in the brown features. The procedure would be significantly complicated by the bright reflections from the surface of each candy.

Thresholding color images requires that the user understand the various color spaces in which the image can be represented, as discussed in the preceding chapter, and what distinguishing color characteristics of the features of interest have. The individual RGB or HSI channel histograms are not much help because they contain no information about the combination of color values for the various pixels. Color space is three dimensional, and so some kind of three-dimensional histogram presentation is very useful. One way to do this is shown in Figure 3.67. The HSI space is modeled in this illustration as a cylinder. The circular cross-section of that cylinder is the Hue-Saturation wheel, in which hue is the angle and saturation is the radius. The fraction of pixels in the image with each combination of hue and saturation is shown as the intensity of the corresponding point on the circle. The axis of the cylinder is the intensity scale, with a conventional histogram display.

Clusters of points in the hue-saturation plane typically indicate the presence of specific colors in the image. Marking a region in this plane, and setting limits on the intensity histogram, selects the ranges of values for the threshold. The requirement that the pixel values must meet all three requirements (the Boolean AND) is automatic and produces an interactive display on the image. In the example in the figure, the cluster of points between cyan and blue, with a range of saturations (representing the reflections from the shiny candy surfaces) correspond to the blue candies. The thresholded image shows the selection.

Using a three-dimensional color space threshold is not only useful as a tool, it is also a good way to learn about color representations and to educate the eye to recognize the components that distinguish colored features. The display contains other information as well. Note that the various colors are represented in the hue-saturation wheel by spots that are generally elongated in the radial (saturation) direction. This corresponds to the variation in color that results from the shiny reflections

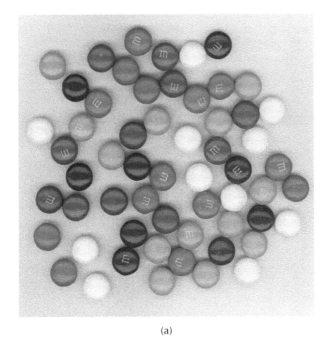

(a)

(b)

FIGURE 3.67 Color thresholding in three dimensions: (a) image of sugar-coated chocolate candies from a desktop flatbed scanner (see color insert following page 150); (b) selection of just the blue candies; (c) the histogram display, with the hue-saturation wheel representing a projection along the axis of the HSI cylinder and the histogram showing the variation in intensity. The limits shown select the blue candies.

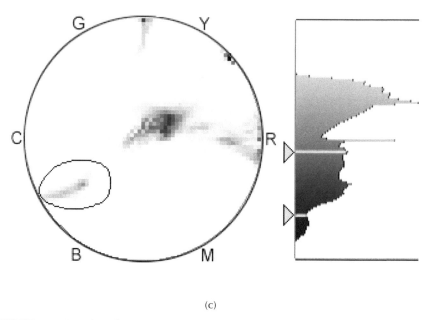

(c)

FIGURE 3.67 (continued)

from the surface, produced by the light source in the flat-bed scanner used to acquire the image. The presence (and length) of the tail is a direct measure of that shininess. The shininess of the sugar candy surface is an important consumer criterion for these candies (I have been told that candies that do not come out sufficiently shiny are not stamped with the little "m" and are sold unbranded through discount houses.)

Interactive threshold setting can be performed by a process of setting limits on a histogram display while observing the results on the image. This generally requires the user to develop some understanding of color space representation. Another approach relies upon the user to recognize the color(s) of interest, but not to interpret them in any way. Instead, the user marks a point or small region on the image that has the color of interest. Then the program finds all of the other pixels that are close to that color. In most cases the additional restriction is made that the selected pixels must be connected to the point original chosen by a continuous path through pixels within the same color range. This is very useful for selecting individual objects (such as the candies in the preceding image). Once the user is satisfied with the initial color selection, it is simple to have the program select all of the pixels in the image that lie within the same color range.

Figure 3.68 shows an example. The color of cheese on a cooked pizza is used to measure the area that has browned in the oven, a consumer quality parameter. The arrow on Figure 3.68a shows the original selection and the initial region that is connected to it and similar in color. Figure 68b shows the selection of all pixels on the image that fall within the same color range. The selected region can then be measured to determine the total area fraction of the pizza that is not browned and the maximum dimension (width) of the unbrowned area.

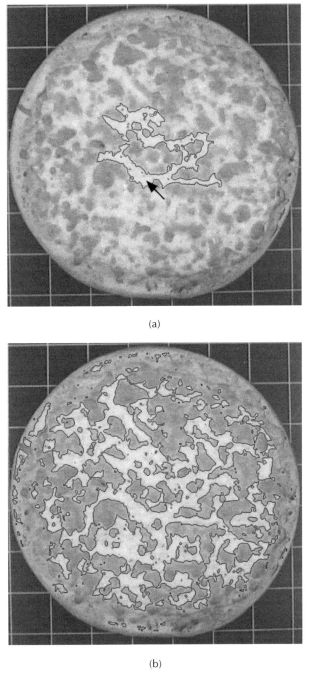

(a)

(b)

FIGURE 3.68 (See color insert following page 150.) Region growing: (a) image of a cooked pizza with the region selected by clicking at the marked location and including all contiguous pixels with RGB values within a tolerance of 32 of the initial point; (b) selection of all similar pixels in the image.

Note that in this type of procedure the region-growing procedure is iterative. Starting at the initial selection point each neighboring pixel is examined and either added to the growing region or not, depending on the color match. If the pixel is added, then its neighbors must be similarly examined and so on, until no more candidates are found. This is in many ways the exact opposite of the top-down split-and-merge approach mentioned above. One problem that this method faces is the unfortunate use of the name "magic wand" to describe the process of clicking at a point on the image and having a selected region appear. Of course, it is not magic and would be better described in technical publications as a region-growing selection based on a user-selected color value.

The principal real difficulty with the region-growing approach lies in deciding whether the colors are matched. Usually this is done in RGB space (because the image is stored that way in the computer) with a single parameter, the difference in intensity value in each channel between the initial selected point and the tested pixel. That amounts to thresholding by selecting pixels whose color values lie within a small cube in RGB space.

Much greater flexibility can be achieved by allowing for different tolerances in each channel, and using HSI channels rather than RGB. Also, it is possible to test pixel similarities against their immediate neighbors rather than the initial value, which makes it possible to accommodate gradual changes or shading across the image. Both of these modifications increase the complexity of the procedure.

MANUAL MARKING

Many of the examples of thresholding methods described above have included varying amounts of human interaction and selection. The automatic methods generally try to minimize that involvement because it can introduce person-to-person and time-to-time variability, but even with the automatic techniques the selection of an appropriate algorithm depends upon the user's independent knowledge about the nature of the sample and the image. Some thresholding tasks can be quite complex, and the next chapter covers the various processing operations that often follow thresholding to correct for inadequacies in the procedures. There are certainly situations in which simple and direct human selection of the features and dimensions to be measured is the fastest and most appropriate method to be used, especially with a small number of images.

Manual measurement of images using rulers, dividers, circle templates, and the like has been performed since the beginning of image quantification. Many of the stereological procedures described in Chapter 1 involve placing various kinds of grids onto the image and counting the hits that the grid points or lines make with the structure of interest, and this, too, has often been performed manually. But people are not very good at measurement or counting. The errors they make have been documented many times, and there is great attraction in using the computer to do what it does best (measurement, counting, recording of results) while retaining the human ability for recognition of subtle details.

No discussion of thresholding would be complete without allowing for this human involvement. One of the most efficient ways to accommodate it is to have the computer acquire the image and perform whatever processing is needed to improve the visibility of the structures of interest. The drawing of appropriate grids on the image is also easy with computer graphics. Then allow the human to mark the points where the grid touches the features of interest. If this is done in a contrasting color, it provides a permanent record of what was counted, and the computer can then easily count the marks. In many cases, the use of black or white marks on a color image, or of bright red, green, etc. on a grey scale image, allows for easy marking of points or drawing of lines.

For measurement purposes, it is practical in some situations to have the user outline the feature boundaries (this is usually easier with computer peripherals such as a tablet and pen rather than a computer mouse, but that may be a matter of personal preference). But if a measure of feature size is desired, the user may employ circle-drawing or line-drawing tools to make simplified representations on the image which the computer can then measure. In Figure 3.69, the maximum dimension of the pore intersections seen in a bread slice can be marked by drawing straight lines on the image. The length and angle of those lines can then be determined and recorded by the computer to characterize the structure.

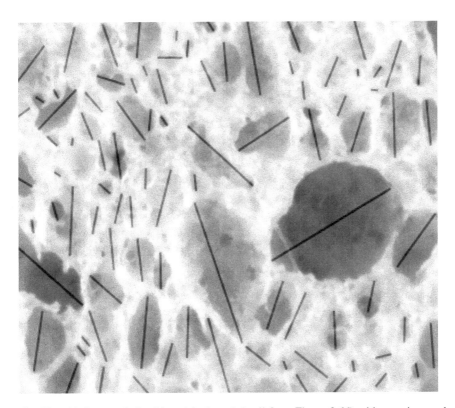

FIGURE 3.69 Image of sliced bread (enlarged detail from Figure 3.66) with superimposed user-drawn lines along the major axis of each pore.

SUMMARY

Processing of images is performed using global procedures that affect all pixels in the same way, or local ones that compare each pixel to its local neighborhood, or procedures that use Fourier space that treat the image as being made up from a collection of sinusoidal ripples. Processing is intended to improve the visibility or detectability of some component of the image. This is usually accompanied by suppression of other components or information.

Human vision has limited ability to discern small variations in brightness. Manipulation of the contrast function, and local sharpening by increasing local brightness changes (sharpening, unsharp masking, and difference of Gaussian processing), can improve the visibility of fine detail. The use of false color and surface rendering also helps the visibility and comparison of fine structures.

Many images are characterized by edges of features, and it is those boundaries that often define the structure that will be measured. Finding edges and steps comprises a major area of image processing algorithms, with many techniques based on intensity gradients, multiple convolutions, statistical values, and other methods. Texture and orientation also play a role in human discrimination of structure, and operators have been developed that can locate and differentiate texture or orientation and convert them to grey scale values to allow thresholding.

The top-hat filter is a powerful tool for locating features of a known size, and is especially useful for locating spikes in Fourier power spectra. The power spectrum also facilitates the measurement of structural periodicities. Cross-correlation using Fourier methods is useful to locate features, and auto-correlation identifies the average size of structuring elements. Cross-correlation is also used to align images so that they can be combined arithmetically. Differences, ratios, and other combinations provide ways to compare or merge information from several images.

The purpose of processing may be simply to improve visibility of image detail, but in the context of image measurement is a precursor to image thresholding. Thresholding selects a range of brightness or color values that correspond to the objects or structures of interest, to isolate them from the rest of the image. Automatic setting of thresholds is preferred to manual methods because it is more reproducible and permits automation, but selecting an appropriate criterion and algorithm requires knowledge about the nature of the sample and the image acquisition process.

In some cases, manual marking of the image dimensions and features to be measured or counted is a more efficient procedure than processing and thresholding. In either case, the intent is to produce a binary (black and white) image that delineates the important structural components and features. The thresholded images are rarely perfect, and further processing of the binary image both to correct problems that may be present and to begin extracting the important measurement parameters is covered in the next chapter.

4 Binary Images

After images have been thresholded, converting the original color or grey scale values to black and white (as a reminder, the convention used in this text shows black features on a white background), there are still processing operations that can be applied. These may include erosion and dilation to clean up imperfect delineation of features in the thresholding procedure, segmentation to separate touching features, or Boolean logic to combine several different thresholded images (such as different color channels) to select the pixels of interest. In addition, several of the binary processing routines convert the original image to one in which the desired measurement information becomes more accessible. Examples include skeletonization, which simplifies shapes to extract basic topological information, and the application of various grids, which allow automatic counting to perform various stereological measurements. Binary image operations are generally separated into two groups: Boolean operations combine multiple images in order to select pixels or objects according to multiple criteria, while morphological operations operate on an image to add or remove pixels dependent on various shape-dependent criteria. Erosion and dilation, skeletonization, and watershed segmentation are all examples of morphological functions.

EROSION AND DILATION

Because of the variation in pixel brightness values in structurally homogeneous regions, thresholding often leaves errors at the pixel level. Particularly around feature boundaries, individual pixels may be added to or missing from features. Erosion (removal of pixels) and dilation (adding pixels) are commonly used to correct for these imperfections. Erosion and dilation were introduced in connection with rank neighborhood operations on grey scale images, in which the pixel was replaced by its lightest (erosion) or darkest (dilation) neighbor. When that same logic is applied to a thresholded binary image, it produces classical erosion in which a black pixel that touches any white pixel as an immediate neighbor is set to white, and conversely dilation in which any white pixel that touches a black pixel is set to black.

This basic operation is often modified in the case of binary images to provide more control over the addition or removal of pixels. One such modification is to count the number of pixels of the opposite color that touch, and to set a minimum number of opposite colored neighbors that must be present to cause a pixel to flip. For instance, if more than four out of the eight adjacent pixel positions are occupied by pixels of the opposite color, the operation functions identically to a median filter. It is also possible in some cases to specify particular patterns of neighbors that will result in erosion or dilation.

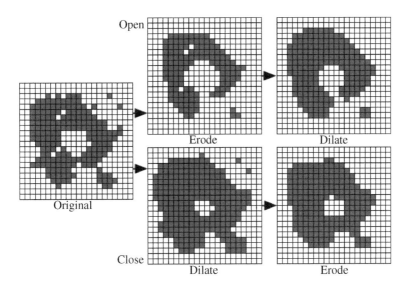

FIGURE 4.1 Example of the application of an opening (erosion followed by dilation) and a closing (dilation followed by erosion), both affecting pixels with more than two neighbors of the opposite color.

Erosion reduces the size of features by removing pixels, while dilation does the reverse. It is very common to use these processes together so that the final feature area is kept about the same. The sequence of an erosion followed by a dilation is called an opening, and the opposite sequence of dilation followed by erosion is called a closing, as shown in Figure 4.1. Notice that the two procedures do not give the same result. In addition to smoothing feature boundaries and removing isolated pixel noise, the opening tends to open up gaps where features touch while the closing tends to fill in breaks in features. Choosing which is correct in any given instance depends on knowing something about how the specimen was prepared and how the image was acquired.

These operations are called "morphological" procedures because they alter the shapes of features, generally producing smoother boundaries but also changing the shape in other ways as will be discussed below. Also, the selection of a proper erosion or dilation procedure generally assumes some knowledge on the part of the user of the true feature shape. In Figure 4.2 an opening was used to remove the pixels that were included in the thresholded foreground because the amount of stain in the tissue was enough to darken the region. It is common with staining to find that some of the dense stain finds its way to places that should not be included, so the selected procedure in this case is to perform an opening, first eroding and then dilating the image. The opposite selection, a closing, would be used to fill in gaps due to cracks in features, for example.

In Figure 4.2, the first step in the closing is an erosion to remove pixels with more than two neighbors of the opposite color. This process was repeated six times,

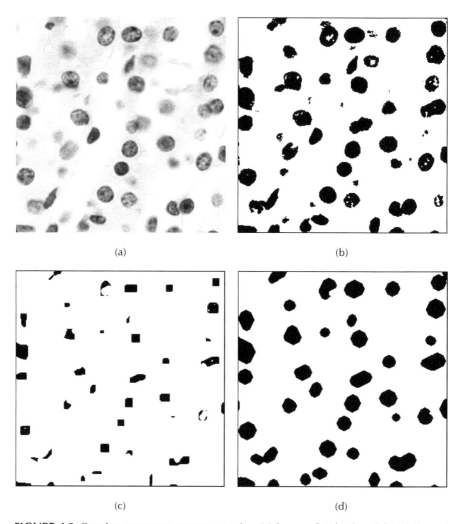

(a)   (b)

(c)   (d)

FIGURE 4.2 Opening to remove extraneous noise: (a) image of stained nuclei (red channel from a color original image); (b) thresholded; (c) eroded; (d) dilated (erosion + dilation = opening). The criteria used were a neighbor count greater than two and six iterations.

until the thin or small features, which are judged not to represent actual features of interest, have been removed. The remaining features are now too small, so applying a dilation to add pixels around the edges, using the same neighbor criterion and number of iterations for a dilation adds back pixels to bring the areas of the resulting dark regions to about the same size as the original, but of course features that were removed entirely do not grow back. Notice also that the shapes of the features that remain have been altered.

Adjusting the required neighbor count provides a very fine level of control over what detail can be selectively kept and what eliminated. The drawing in Figure 4.3

(a)

(b)

FIGURE 4.3 Drawing of a panda: (a) original; (b) closing, neighbor count > 0, 1 iteration; (c) opening, neighbor count > 0, 1 iteration; (d) dilation, neighbor count > 7, 1 iteration; 2) erosion, neighbor count > 7, 1 iteration.

(c)

(d)

FIGURE 4.3 (continued)

(e)

FIGURE 4.3 (continued)

illustrates some of the possibilities. It contains isolated pixel noise (shot noise) in both the white and black areas, as well as fine lines and narrow gaps. Applying a single iteration of classical open or closing can be used to fill in the gaps (closing) or remove the fine lines (opening). In classical erosion and dilation, each pixel with any touching neighbors of the opposite color is flipped to that color. By changing only those pixels that have more than seven (i.e., all eight touching neighbors) adjacent pixels of the opposite color, it is possible to use erosion and dilation to selectively remove just the isolated pixel noise, without altering anything else in the image.

While it is often possible to remove noise from an image by using erosion and dilation after thresholding, in general this is not as good a method as performing appropriate noise reduction on the grey scale image before thresholding. More information is available in the grey scale image to reduce the noise, using tools such as the median filter or maximum likelihood filter to evaluate the grey scale values in the neighborhood and adjust pixel values accordingly. As shown in Figure 4.4, this makes it possible to preserve fine detail and local boundary shapes better than applying morphological operations to the binary image after thresholding.

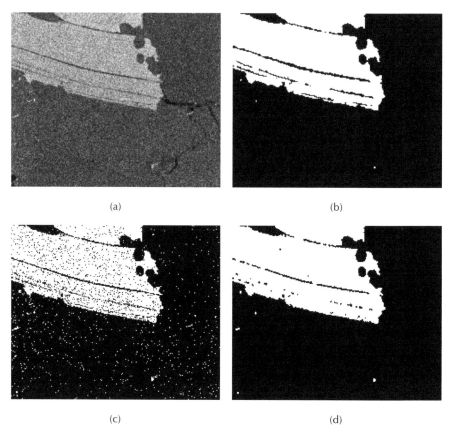

(a)   (b)

(c)   (d)

FIGURE 4.4 Noise reduction: (a) original X-ray map; (b) grey scale noise reduction before thresholding; (c) thresholding applied to image (a); (d) closing (neighbor count > 5, 1 iteration) applied to image (c).

THE EUCLIDEAN DISTANCE MAP

The shape modifications introduced by erosion and dilation are of two kinds: those that are intended and which are ideally dependent only on the shape of the original thresholded feature, and those that occur because of the nature of the square pixel grid from which the features are constructed. The latter alterations are unfortunately not isotropic, and alter shapes differently in horizontal and vertical directions as compared to 45 degree diagonal directions. Figure 4.5 shows a circle. Ideally, erosion and dilation would act on this circle to produce smaller or larger circles, with a difference in radius equal to the number of iterations (pixels added or removed).

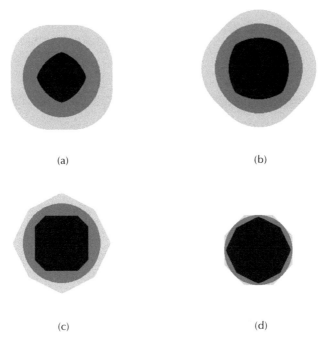

(a)   (b)

(c)   (d)

FIGURE 4.5 Anisotropy of erosion/dilation (grey circle eroded 25 times to produce the black inner shape, dilated 25 times to produce the light grey outer shape: (a) classic; (b) neighbor count > 1; (c) > 2; (d) > 3.

In fact, this is not the actual result. The dark figures show the effect of erosion using twenty-five iterations. When classic erosion is used, pixels are removed that touch any background pixel. Removing a pixel in the diagonal directions changes the dimension by the diagonal of the square pixel, which is 41% greater than the width of the pixel in the horizontal or vertical directions. Hence, the erosion proceeds more rapidly in the diagonal directions and the original circle erodes toward the shape of a diamond. Similarly, the light grey figure shows the result of dilation, in which the feature grows more in the diagonal direction and evolves toward the shape of a square.

Using other neighborhood criteria can alter the shape of the eroded or dilated features, but it cannot produce an isotropic circle as a result. The figure shows the effects of requiring more than 1, more than 2 or more than 3 neighbor pixels to be of the opposite color. The shapes erode or dilate toward the shapes of squares, diamonds or octagons, but not circles. Also the size of the resulting shapes varies, so that the number of iterations cannot be used directly as a measure of the size change of the feature. These problems limit the usefulness of erosion and dilation based on neighboring pixels, and in most cases it is not wise to use them for the addition or removal of more than 1 or 2 pixels.

There is another approach to erosion and dilation that solves the problem of anisotropy. The Euclidean distance map (EDM) is used with binary images to assign

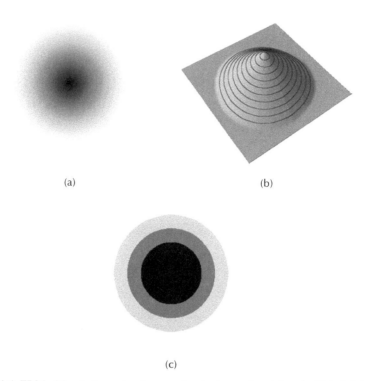

(a)   (b)

(c)

FIGURE 4.6 EDM of the circle, rendered as a surface to show values (plotted with isobrightness contour lines) and isotropic erosion/dilation using the EDM (compare to preceding image).

to every pixel that is black (part of the thresholded structure or features) a number that measures the distance from that pixel to the nearest white (background) pixel, regardless of direction. The construction of this grey scale image is very rapid, and as shown in Figure 4.6 produces values that can be rendered as though they were a surface to produce a smooth mountain with sides of uniform slope. Thresholding this grey scale image is equivalent to eroding the original binary image. To perform dilation, the image is inverted and the EDM is computed for the background, and then thresholded. As shown in the figure, erosion and dilation using the EDM method is isotropic and produces larger or smaller features that preserve shape. Because it is not iterative, the method is also much faster than the traditional method.

In general, the EDM-based erosion and dilation methods should be used whenever more than 1 or 2 pixels are to be removed from or added to feature boundaries, because of the improved isotropy and avoidance of iteration. However, the traditional method using neighbor relationships offers greater control over the selective removal of (or filling in of) fine lines and points. Also, the EDM-based method requires that the assigned grey scale pixels values that represent distance have more precision than the integer values that grey scale images usually contain. The distance between a feature pixel and the nearest point in the background is a real number whose precision must be maintained for accurate results, and not all software systems do

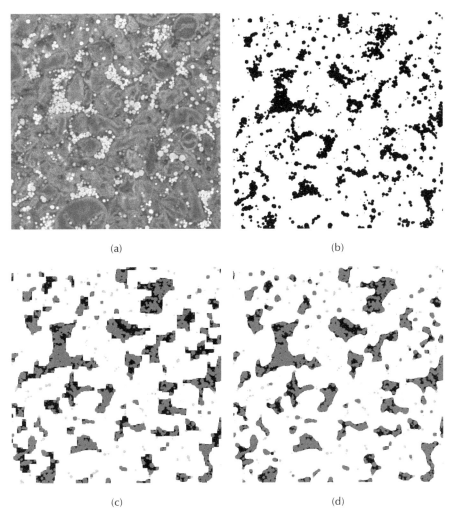

(a)   (b)

(c)   (d)

FIGURE 4.7 Structural phase separation in custard: (a) grey scale rendering of original color image (detail of Color Figure 4.7, nile red stain applied to fat globules; see color insert following page 150); (b) fat droplets thresholded from red channel; (c) classic closing (six iterations) and opening (three iterations) to delineate phase regions; (d) EDM-based closing and opening, distance of 6.5 pixels. Images (c) and (d) are shown with original thresholded image overlaid for comparison; the dark grey areas are the original thresholded pixels, the light grey areas were removed by the opening, the black regions filled in by the closing).

this properly. This becomes particularly important for some of the additional uses of the EDM discussed below.

Just as for traditional morphological erosion and dilation, the EDM-based operations can be combined into openings and closings. In many cases, both are needed. For instance, Figure 4.7 shows the distribution of stained fat particles in custards. It is the clusters of particles that are important in determining the rheology, appearance, and other properties of the product. Simply thresholding the red channel of the original color image produces a good binary image of the distribution of nile red stain in the sample. But there are gaps within the cluster regions that must be filled in, and a few isolated particles in the other portions of the structure that should be erased.

Applying a closing to fill in the gaps, and an opening to remove the isolated, scattered features, produces an image in which the total volume fraction and size of the clusters and the spacing between them can be measured. However, if done with classic morphological methods this process produces suspicious shapes in which 90- and 45-degree boundary lines are prevalent. The EDM-based erosion and dilation procedures generate shapes that are much more realistic and better represent the actual structure present, so that more accurate measurements can be made.

SEPARATING TOUCHING FEATURES

The Euclidean distance map has other uses as well. Watershed segmentation uses the EDM as the basis for separating touching, convex shapes. This is primarily useful for section images or surface images in which features are adjacent or overlap only slightly. It is less useful for arrays of three-dimensional particles because particles of different sizes may overlap significantly, and small particles may be hidden by larger ones. The method also has difficulties with jagged or irregular shapes, and depending on the quality of the EDM used to control the separation, may cause the breakup of fibers.

Figure 4.8 shows the basic logic behind watershed segmentation. Thresholding the image of touching particles produces a binary representation in which all of the pixels touch, and would be considered to be part of a single feature. The grey scale values generated by the EDM, when represented as a rendered surface, produce a mountain peak for each of the particles. Imagine that rain falls onto this terrain; drops that strike the mountains run downhill until they reach the sea (the background). Locations on the saddles between the mountains are reached by drops running down from two different mountain peaks. Removing those points separates the peaks, and correspondingly separates the binary images of the various particles as shown in the figure. Notice that some boundaries around the edge of the image are not separated, because the information is missing that would be required to define the mountain peak beyond the edge. The edge-touching features cannot be measured anyway, as discussed in the chapter on feature measurements, so this does not usually create analysis difficulties.

The watershed method works well when particles are convex, and particularly when they are spherical (producing circular image representations). It happens that in many cases, due to membranes or physical forces such as surface tension, particles are fairly spherical in shape and, as shown in Figure 4.9, the separation of the particles for counting or individual measurement can use the watershed segmentation method.

There is an older technique that is also used with spherical particles that has found its principal application in fields other than food science, but which can certainly be used for food structure images as shown in Figure 4.10. The two methods for erosion and dilation described earlier in this chapter apply to binary images after thresholding. A third erosion/dilation procedure was described in an earlier chapter that is applied to grey scale images. Used, for instance, to remove features from an

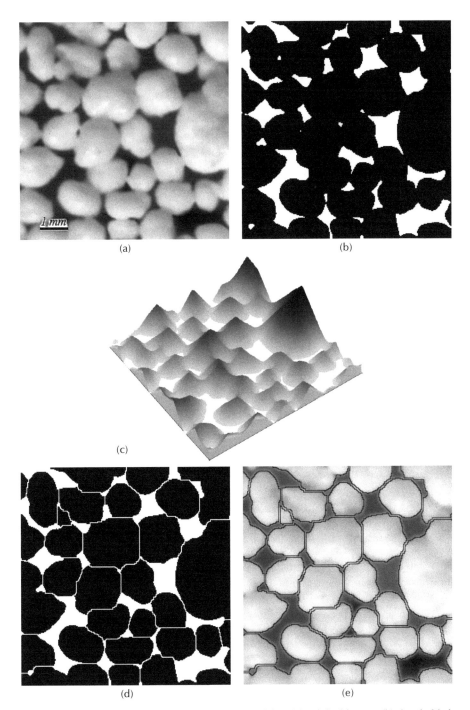

FIGURE 4.8 Watershed separation of touching particles: (a) original image; (b) thresholded; (c) EDM of thresholded image rendered as a surface to show mountain peaks, passes, and background; (d) watershed separation; (e) feature outlines superimposed on the original image.

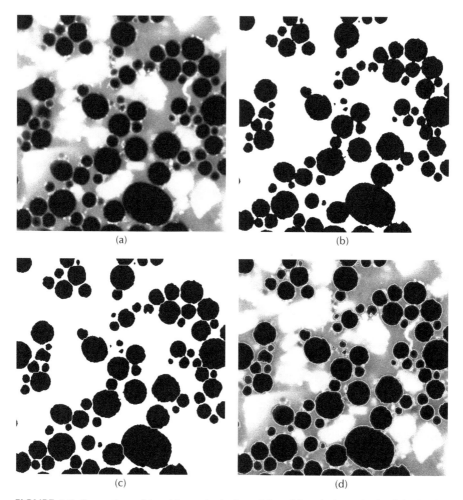

FIGURE 4.9 Separation of touching spherical particles: (a) red channel of original color image of mayonnaise; (b) thresholded fat globules; (c) watershed segmentation of (b); (d) overlay of separate particle outlines on original image.

image and produce a background that can be used to level an image by subtraction or division, grey scale erosion and dilation replace each pixel with the brightest or darkest (respectively) pixel values in a neighborhood.

When applied to an image such as the one shown in the figure, the process shrinks all of the features by one pixel on all sides. If the particles are convex and not overlapped by more than 50%, they will eventually separate before disappearing completely as the procedure is repeated. By counting the number of features that disappear at each iteration of erosion or dilation, a distribution of particle sizes can be obtained. Since each iteration reduces the radius of the particles by 1 pixel, the distribution is the number of features as a function of the inscribed radius. Performing the counting operation requires some additional Boolean logic that has not yet been introduced, but is relatively straightforward and very fast.

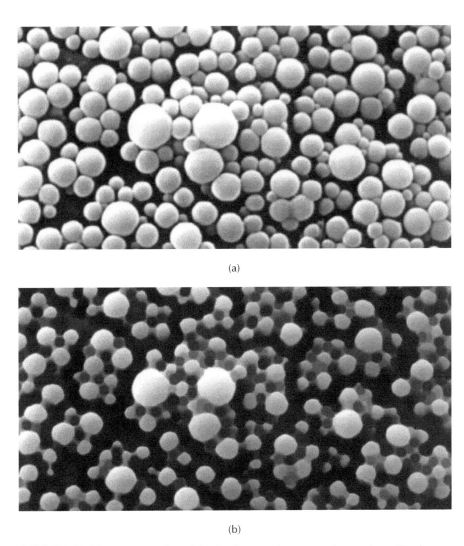

(a)

(b)

FIGURE 4.10 Measurement of particle size by counting successive erosions. The images here show four stages of successive grey scale erosion of the features (each image has the particles reduced in radius by four pixels). Overlapped and touching features separate before their final disappearance. Counting the number of features that disappear after each iteration provides a size distribution.

The method of successive erosion and counting is iterative and slow, and can be replaced by an equivalent procedure using the EDM. In the example in Figure 4.8(c), the peaks of the mountains correspond to the various features present, and the height of each mountain (the value of the EDM at the maximum point) measures the radius of an inscribed circle in the feature. It is not necessary to actually perform the watershed segmentation to measure this size distribution. Simply finding each local maximum in the EDM and recording their values provides a measurement of

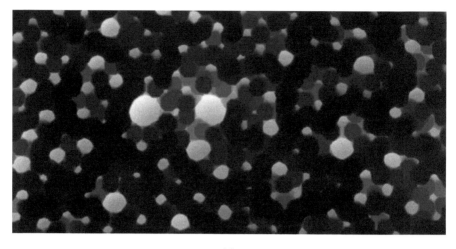

(c)

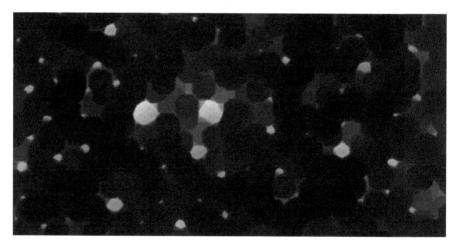

(d)

FIGURE 4.10 (continued)

the size (specifically, the radius of an inscribed circle) for each touching or over-lapped feature. These local maxima in the EDM are called the "ultimate eroded points" to emphasize that they are the same points that would be the last pixels within each feature to remain in the iterative erosion procedure shown above. However, they have the advantage that the EDM method is isotropic, whereas iterative erosions, as noted above, can introduce directional bias. In Figure 4.11, the ultimate eroded points are marked on the touching spherical particles from Figure 4.9.

Watershed segmentation cannot always successfully separate touching objects. In Figure 4.12, thresholding the image of cells produces a binary image in which the cells have holes within them, the result of low density vacuoles in the grey scale image. Attempting to apply a watershed procedure to this image breaks the cells up

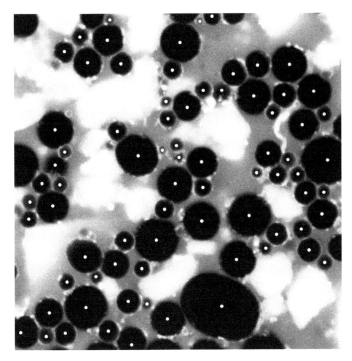

FIGURE 4.11 The ultimate eroded points (UEP) are local maxima in the EDM, marked on the original image from Figure 4.9. The value of the EDM at the ultimate point is a direct measure of particle size without requiring watershed segmentation, corresponding to the radius of an inscribed circle in each feature.

into pieces as watershed lines are drawn to the interior holes (Figure 4.12c). Simply filling all of the holes will not correct the problem, because the image also contains holes that lie between the cells. However, the size and shape of the interior holes are different from those of the holes between cells: the latter are larger and/or cusp-shaped while the interior holes are small and/or round. Selecting just the interior holes and filling them in allows a watershed procedure to successfully separate the touching cells (Figure 12e). The selection process is performed by measurement of size and shape, as discussed in Chapter 5, after which a Boolean OR is used to combine the image of the selected interior holes with the original binary image.

Angular features can also create difficulties for the watershed method. As shown in Figure 4.13, when irregularly shaped particles touch the junctions are frequently not simple. The watershed method sometimes draws the separation lines in ways that alter the apparent shape of the features. It may also fail to draw some of the lines where the width of the touching junction is greater than the perpendicular width of the feature. An example is shown by the arrow in Figure 4.13(c). Another problem with this particular image is that the size range of the features is too great for measurement at a single magnification. The shapes of the smallest features (under about 40 pixels) are not adequately represented in this image, and should be eliminated before measurement. Additional images at higher magnification can then be

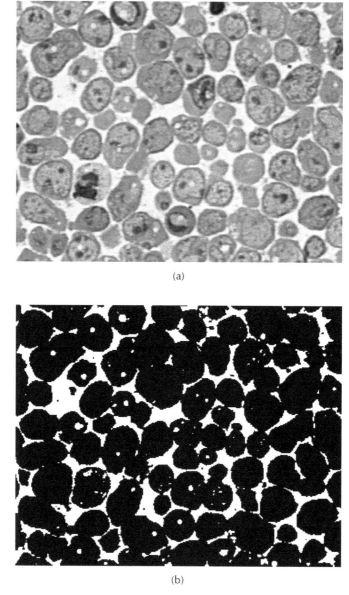

(a)

(b)

FIGURE 4.12 Segmentation of touching features containing holes: (a) original image of cells; (b) thresholded; (c) watershed applied to image (b) cuts cells into fragments; (d) filling holes within cells based on size and shape; (e) successful watershed segmentation of image (d); (f) watershed segmented cell outlines superimposed on original image.

used to measure the small features to include in the overall statistics. The removal of features based on size, shape or other parameters is carried out as part of the measurement process, not by erosion which would also alter the shapes of other features present.

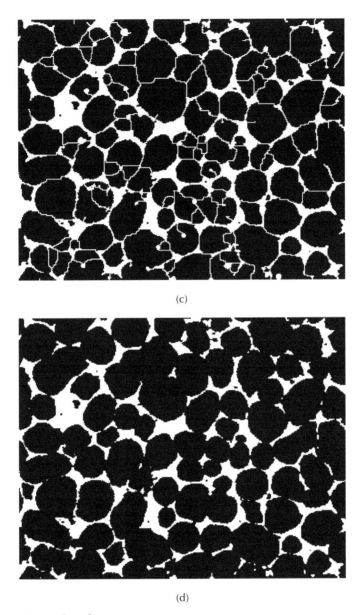

(c)

(d)

FIGURE 4.12 (continued)

Images of fibers or other features with long lines of nearly constant width also create a hazard for the watershed technique. Many programs implement the EDM on which the watershed depends using integer representations of the distance values for each pixel. Because of the aliasing of the pixel image, thin lines are often broken into many small pieces by the watershed, which sees each indentation along the

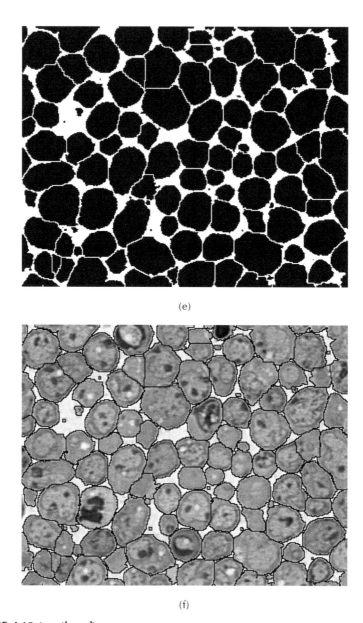

(e)

(f)

FIGURE 4.12 (continued)

edges of the feature (Figure 4.14c and 4.14d). But even a properly constructed EDM in which the distance values have enough precision to accurately reflect the pixel distances will cut fibers where they cross, as shown in Figure 4.14(b). This may or may not be useful. Other ways to segment and measure fiber images, based on the feature skeleton, will be shown below.

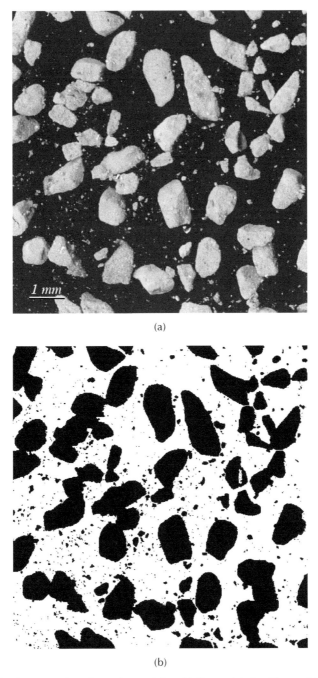

FIGURE 4.13 Segmentation of angular particles: (a) freeze-dried coffee particles (see color insert following page 150); (b) thresholded; (c) watershed segmentation (arrow marks a location where the width of the touching junction is greater than the width of the smaller feature, and no separation line is drawn); (d) removal of features with area less than 40 pixels.

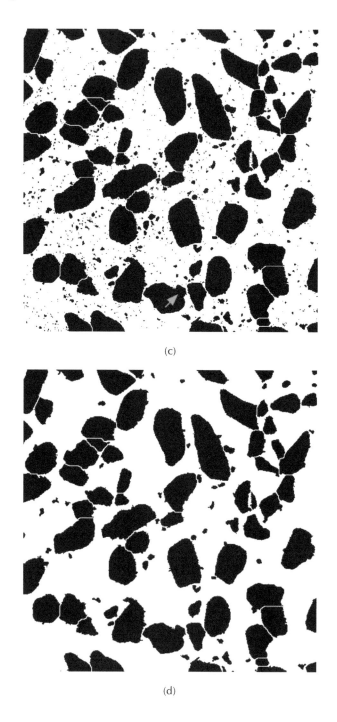

(c)

(d)

FIGURE 4.13 (continued)

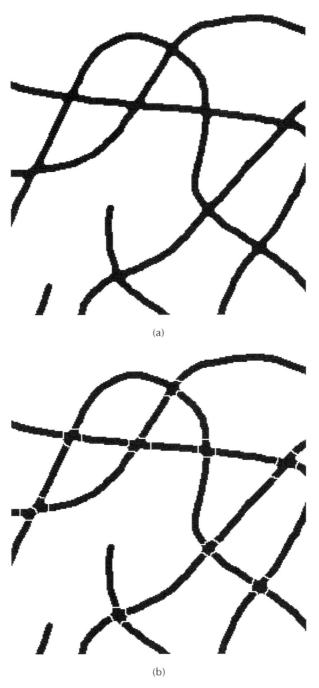

(a)

(b)

FIGURE 4.14 Watershed segmentation applied to an image of crossing fibers: (a) thresholded original; (b) watershed cuts features near crossing points; (c) Image Pro Plus result; (d) NIH-Image result. These two widely used programs implement the EDM with integer arithmetic which creates many erroneous cuts across the fiber images.

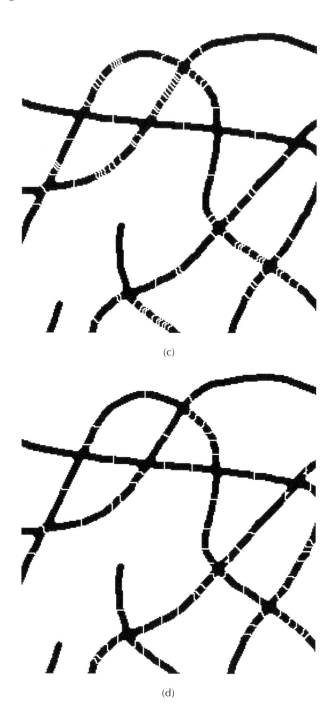

(c)

(d)

FIGURE 4.14 (continued)

BOOLEAN COMBINATIONS

Boolean combinations of two or more binary images of the same scene can be used to isolate specific structures or features. The use of a Boolean AND has already been introduced in the context of color thresholding, to combine multiple criteria from the various color channels. AND keeps just the pixels that are selected in both of the source images. The other principal Boolean relationships are OR, which keeps pixels selected in either of the source images, and Exclusive-OR (Ex-OR) which keeps those pixels that are different, in other words pixels that are selected in either image but not the other. These relationships are summarized in Figure 4.15, and will remind many readers of the Venn diagrams they encountered in high school.

These three Boolean operators are frequently used with the logical operator NOT. This inverts the image, making the black pixels white and vice versa, or in other words selects the pixels which had not been selected before. Arbitrarily complex combinations of these operations can be devised to make selections. But it is important to use parentheses to explicitly define the order of operations. As Figure 4.16 shows, a relationship such as A AND (NOT B) is entirely different from NOT (A AND B). This becomes increasingly important as the complexity of the logic increases.

In addition to the use of Boolean operations to combine thresholding criteria for various color channels, there is a very similar use for images acquired using different wavelength excitation or filters, or different analytical signals such as the X-ray

(a)   (b)   (c)

(d)   (e)

FIGURE 4.15 Boolean combinations of pixels: (a, b) original sets; (c) A AND B; (d) A OR B; (e) A Ex-OR B.

FIGURE 4.16 Combinations of Boolean operations: (a) A AND (NOT B); (b) NOT (A AND B); (c) (NOT A) AND B; (d) (NOT A) OR B; (e) NOT (A OR B).

maps from an SEM. But the same procedures can also be applied to different images that have been derived from the same original image in order to isolate different characteristics of the structure. Figure 4.17 shows an illustration. The original figure was created with regions having different mean brightnesses and different textures.

To select a region based on both criteria, two copies of the image are processed differently. In one, the texture is eliminated using a median filter, so that only the brightness differences remain, and can be used for thresholding. Of course, it is not possible to completely isolate a desired region based on only this one criterion. Processing the second copy of the image based on the texture produces an image that can be thresholded to distinguish the regions with different texture values, but ignoring the brightness differences. Combining the two thresholded images with a Boolean AND isolates just the selected region that has the desired brightness and texture properties. Obviously, this can be extended to include any of the image characteristics that can be extracted by processing as discussed in the previous chapter.

Boolean logic is also useful for the selection of features present in an image based on their position relative to some other structure. Figure 4.18 illustrates this for the case of immunogold labeling. Most of the particles adhere to organelles but some are present in the cytoplasm. Using two copies of the image subjected to different processing, it is possible to select only those particles that lie on the organelles. The full procedure includes background leveling, as shown in a previous chapter, to enable thresholding the particles. It concludes with an AND that eliminates the particles that are not on organelles.

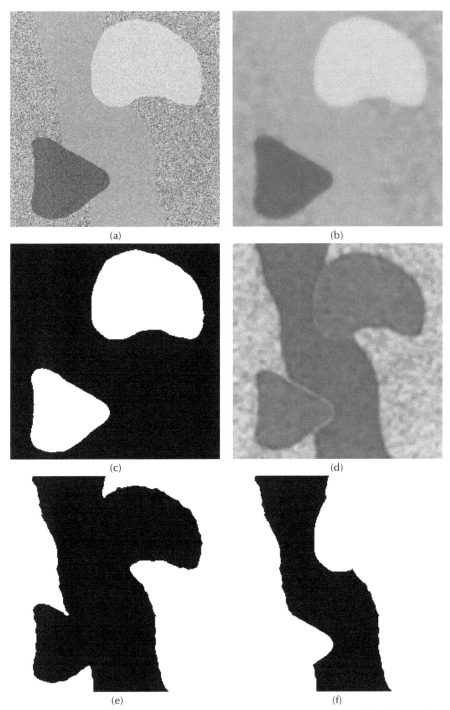

FIGURE 4.17 Selection based on multiple criteria: (a) example of image with regions having different brightness and texture; (b) median filter eliminates texture; (c) thresholded binary from image (b); (d) fractal filter distinguishes smooth from textured regions; (e) thresholded binary from image (e); (f) combining (c) AND (e) isolates medium grey smooth region.

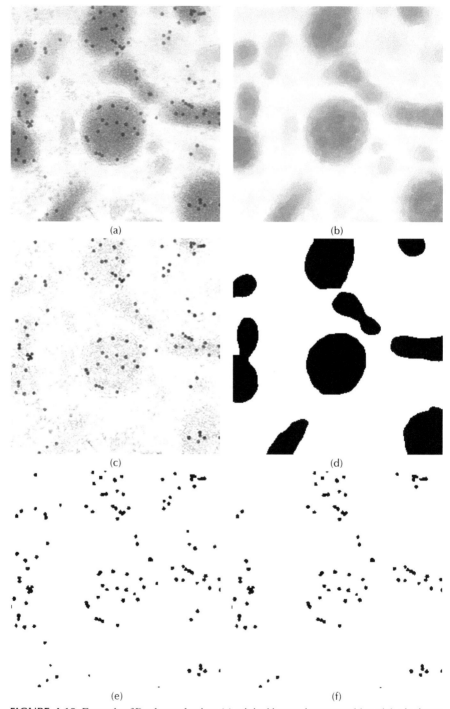

FIGURE 4.18 Example of Boolean selection: (a) original image, immunogold particles in tissue; (b) grey scale opening to remove particles and generate background; (c) (a) divided by (b) levels contrast for gold particles; (d) thresholding (b) gives image of organelles; (e) thresholding (c) gives image of particles; (f) combining (d) AND (e) gives just the particles on the organelles.

Notice that this procedure would not have been able to select the organelles that had gold markers and eliminated any that did not. The simple Boolean operations work on the individual pixels and not on the features that they form. A solution to this in the form of feature-based logic will be introduced below, but first another important use of Boolean logic, the application of measurement grids, must be introduced.

USING GRIDS FOR MEASUREMENT

Many measurements in images, including most of the important stereological procedures, can be carried out by using various types of grids. With the proper grid many tasks are reduced to measuring the lengths of straight line segments, or to counting points. Figure 4.19 shows an example. A section through a coating has

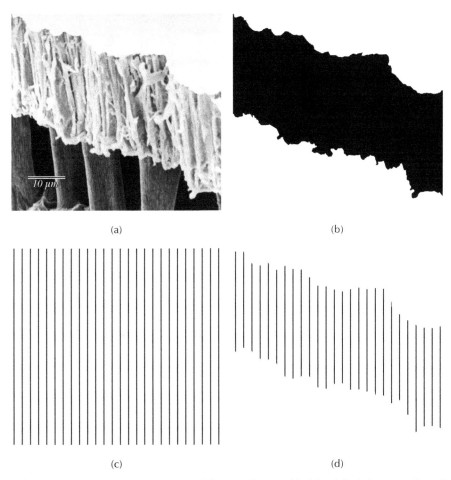

FIGURE 4.19 Measurement of layer thickness using a grid: (a) original (cross-section of sunflower seed hull, original image is Figure 9-8a in Aguilera and Stanley, used by courtesy of the authors); (b) thresholded cross-section; (c) grid of vertical lines; (d) (b) AND (c) produces line segments for measurement (thickness = 21.6 ± 1.3 μm).

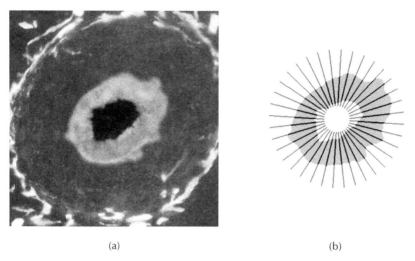

(a)   (b)

FIGURE 4.20 Cross-section through a blood vessel: (a) original; (b) thresholded with superimposed radial line grid.

been prepared and imaged. The thickness of that coating can be described by the minimum, maximum, mean, and standard deviation of the vertical dimension. Thresholding the coating in the image, generating a grid of parallel vertical lines, ANDing the grid with the thresholded binary, and measuring the lengths of the resulting line segments, provides all of that information very efficiently.

Of course, other specific grids can be devised in other situations. For example, a grid of radial lines would be appropriate to measure the thickness of a cylindrical structure, as shown in Figure 4.20. In this particular instance, note that because the vein is not round, but somewhat elliptical (or perhaps the section plane is not perpendicular to the axis of the vein), some of the grid lines intersect the wall at an angle and produce an intersection value that is greater than the actual wall thickness. While this illustrates the use of a radial grid, for this particular measurement there is a better procedure that uses the EDM which will be introduced later in this chapter.

The vertical line grid works in the first example because by definition or assumption it is the vertical direction in the image that corresponds to the desired dimension, and because the section being imaged has been cut in the proper orientation. In the general case of sectioning a structure, however, the cut will not be oriented perpendicular to the layer of interest and the resulting observed width of the intersection will be larger than the actual three-dimensional thickness. Stereology provides an efficient and accurate way to deal with this problem.

As shown in Figure 1.32, if a thick layer of structure exists in three dimensions, random line intersections with the layer (called by stereologists a "muralia") will have a minimum value (the true 3D thickness) only if the intersection is perpendicular to the layer, and can have arbitrarily large values for cases of grazing incidence. A distribution of the frequency of various intercept lengths declines asymptotically for large values. Replotting the data as the frequency distribution of the inverse intercept lengths simplifies the relationship to a straight line. The mean value of the reciprocal

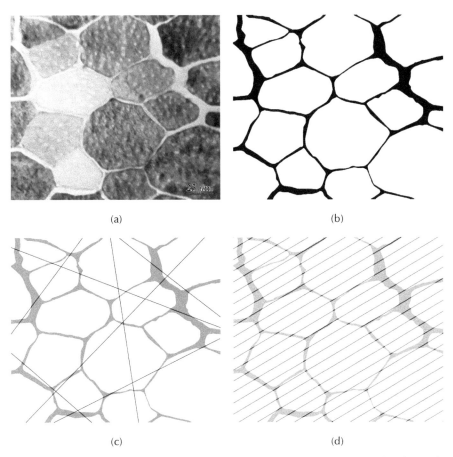

FIGURE 4.21 Measurement of three-dimensional layer thickness: (a) section of pork muscle (reacted for glycogen by periodic acid–Schiff reaction; courtesy of Howard Swatland, Department of Animal and Poultry Science, University of Guelph); (b) thresholded sections through layers; (c) random line grid; (d) parallel line grid in one orientation.

value on the horizontal axis is just two thirds of the maximum, which in turn is the reciprocal of the true three-dimensional thickness.

So the measurement problem reduces to finding the mean value of the reciprocal of the intercept length for random intersections. If the section plane is random in the structure, this could be done with a set of random lines as shown in Figure 4.21(c). Random lines are very inefficient for this purpose, because they tend to oversample some parts of the image and undersample others; a systematic measurement that covers all orientations can be used instead. A grid of parallel lines is generated at some angle, ANDed with the binary image of the layer, and the intersection lengths measured. A running total of the inverse of the intercept values can be stored in memory as the procedure is repeated at many different angles, to produce a robust estimate of the mean.

In the example, the measured mean reciprocal intercept value (from more than 2000 total intercepts obtained with 18 placements of the parallel line grid, in rotational steps of 10 degrees) is 0.27 μm^{-1}. From this, the true mean three-dimensional thickness is calculated as 2.46 μm. Thus, a true three-dimensional thickness of the structure has been measured, even though it may not actually be shown anywhere in the image because of the angles of intersection of the section plane with the structure.

In the preceding examples, it was necessary to measure the length of the intersections of the lines with the features. In many of the stereological procedures shown in Chapter 1 for structural measurement, it is only necessary to count the number of hits, places where the grid lines or points touch the image. This is also done easily by using an AND to combine the grid (usually generated by the computer as needed, but for repetitive work this can also be saved as a separate binary image) with the image, and then counting the features that remain. Note that these may not be single pixel features; even if the intersection covers several pixels it is still counted as a single event.

In the example shown in Figure 4.22, a vertical section is used to measure the surface area per unit volume of bundles of stained muscle tissue. Thresholding the image delineates the volume of the muscle. The surface area is represented by the outline of these regions. The outline can be generated by making a duplicate of the image, eroding it by a single pixel, and combining the result with the original binary using an Exclusive-OR, which keeps just the pixels that have changed (the outline that was removed by the erosion). Generating an appropriate grid, in this case a cycloid since a vertical section was imaged, and combining it with the outlines using a Boolean AND, produces an image in which a count of the features tallies the number of hits for the grid. As noted in Chapter 1, the surface area per unit volume is then calculated as $2 \cdot N/L$ where N is the number of hits and L is the total length of the grid lines.

Automatic counting of hits made by a grid also applies to point grids used to measure area fraction and volume fraction. Figure 4.23 shows an example. The original image is a transverse section through a lung (as a reminder, for volume fraction measurements, vertical sectioning is not required since the points in the grid have no orientation). Processing this image to level the nonuniform brightness enables automatic thresholding of the tissue. Because the image area is larger than the sample, it will also be necessary to measure the total cross-sectional area of the lung. This is done by filling the holes in Figure 4.23(b) to produce Figure 4.23(c) (filling holes is discussed just after this example).

A grid of points was generated by the computer containing a total of 132 points (taking care not to oversample the image by having the points so close together that more than one could fall into the same void in the lung cross section). ANDing this grid with the tissue image (Figure 4.23b) and counting the points that remain (54) measures the area of tissue. ANDing the same grid with the filled cross-section image (c) and counting the points that remain (72) measures the area of the lung. In the figure, these points have been superimposed, enlarged for visibility, and grey-scale coded to

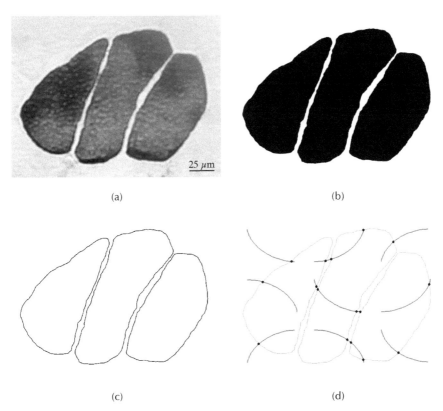

(a)   (b)

(c)   (d)

FIGURE 4.22 Vertical section through pork with slow-twitch muscle fibers stained (acid-stable myosin ATPase reaction, image courtesy of Howard Swatland, Department of Animal and Poultry Science, University of Guelph): (a) red channel from original color image; (b) thresholded binary image of stained region; (c) outlines produced by erosion and Ex-OR; (d) superimposed cycloid grid and intersection points identified with a Boolean AND.

indicate which ones hit tissue. The ratio $(54/72) = 75\%$ estimates the volume fraction of tissue. Of course, to meet the criterion of uniform sampling these point counts should be summed over sections through all parts of the lung before calculating the ratio.

In the preceding example, and several others in this chapter, holes within binary features have been filled. The logic for performing that operation is to invert the image, identify all of the features (connected sets of black pixels) that do not have a connected path to an edge of the image field, and then OR those features with the original image. The grouping together of touching pixels into features and determining whether or not a feature is edge-touching is part of the logic of feature measurement, which is discussed in the next chapter.

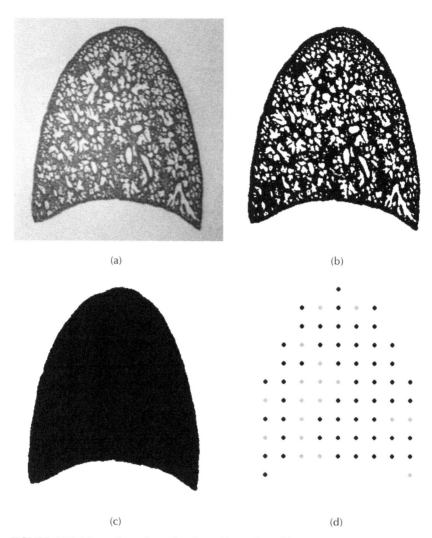

FIGURE 4.23 Measuring volume fraction with a point grid: (a) cross-section of mouse lung tissue; (b) leveled and thresholded; (c) holes filled as described in text; (d) ANDing with a point grid; dark points ANDed with (b), light grey with (c).

Filling of holes within thresholded features is often required when the original image records the outlines of features, as shown in the example of Figure 4.24. The use of edge-enhancing image processing, as described in the preceding chapter, also creates images that, when thresholded, consist of outlines that require filling to delineate features. Macroscopic imaging with incident light sources that produce bright reflections from objects can also require filling of holes.

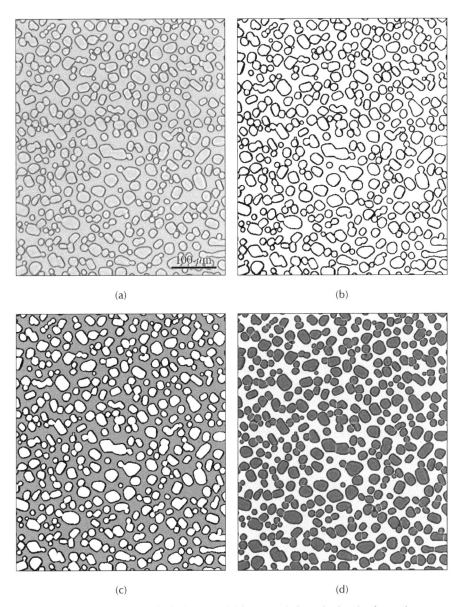

(a)

(b)

(c)

(d)

FIGURE 4.24 Filling holes within features: (a) ice crystals in melted and refrozen ice cream (edges are dark because of light refraction; courtesy of Ken Baker, Ken Baker Associates); (b) thresholded feature outlines; (c) the background (grey) is identified because it reaches the edge(s) of the image field; (d) the filled ice crystals after watershed segmentation (note that a few erroneous separation lines have been drawn due to irregular feature or cluster shapes).

USING MARKERS TO SELECT FEATURES

The examples of Boolean logic shown above operate at the pixel level. At each pixel location, the values in the two source images are compared and the result used to determine the state of the pixel in the derived result, without any regard to any other pixels in the image. For some purposes it is desirable to perform Boolean logic using entire features. This makes it possible to have one image containing markers that identify features of interest, with a second image containing the features themselves, to produce a result in which the features that contain the markers are selected in their entirety.

An example of a Boolean feature-AND is shown in Figure 4.25. Originally a color image, this confocal microscope image shows red cells, some of which contain green nuclei. The challenge is to select those cells that contain the green nuclei as markers. Thresholding the red and green channels, respectively, produces binary images of the cells and the nuclei. Combining these with a pixel-based Boolean AND would produce an image identical to Figure 4.25(d), which is just the marks themselves. A feature-based AND using Figure 4.25(d) as the marker and Figure 4.25(b) as the target instead keeps the entire feature (the cell) if any part of it is selected by the marker (the nucleus).

It is necessary to understand that in the feature-based AND, order is important. A conventional pixel-based Boolean operation commutes (i.e., A AND B produces the same result as B AND A), but the feature-based AND does not. There are several different ways that this process can be implemented. Some systems use a dilation method, in which the markers are dilated iteratively but only pixels selected by the target image can be turned on. The method must be repeated until no further changes occur, and for complex shapes this can take a long time. Another approach first labels all of the features in both images, and if any pixel within a feature is selected by a marker, the entire feature is kept.

The feature-based AND is an important tool, but there is no need for a feature-based OR. That would produce exactly the same result as a pixel-based OR, keeping any feature that was present in either image. A feature-based Ex-OR may be useful in some instances (one is the disector used for stereological counting using two parallel plane images), and may either be implemented directly or by combining a feature-based AND with a pixel-based Ex-OR.

Regardless of the details of the implementation method, the feature-AND technique opens up many possibilities for selecting objects based on the presence of a marker. The marker(s) may in some instances be a set of lines or points used to probe the image, such as a grid or a set of outlines of regions present in the image. Examples of this will be shown below. In many cases the marker is a feature within the object itself, either usually a shape or color tag that can be separately identified. For example, the image in Figure 4.26 shows some candies imaged with a desktop flatbed scanner. Thresholding of the candies is most easily accomplished by selecting the background pixels (a fairly uniform unsaturated grey) rather than the candies, which vary in color. The resulting image has interior holes due to the reflections from the shiny candy surface and the printed "m" characters, but these holes can be filled as described previously. The resulting features touch, but are convex and easily separated by a watershed segmentation.

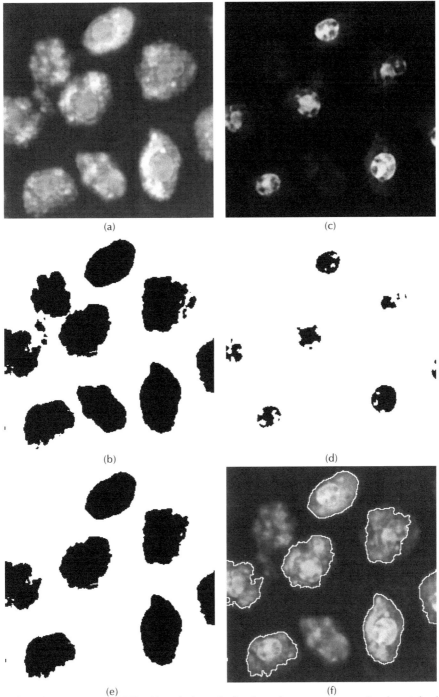

FIGURE 4.25 Feature-AND: (a) red channel of color microscope image showing stained cells; (b) thresholded cells; (c) green channel of same image showing stained nuclei in some of the cells; (d) thresholded nuclei; (e) Feature-AND using (d) as the marker image to select features from (b); (f) outline of selected cells which contain stained nuclei.

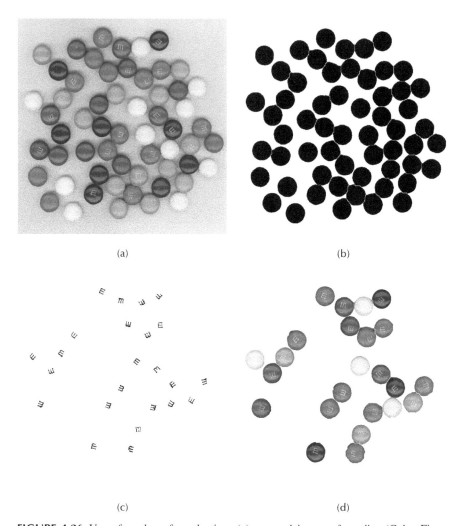

(a)   (b)

(c)   (d)

FIGURE 4.26 Use of markers for selection: (a) scanned image of candies (Color Figure 4.26a); (b) thresholded features as described in text; (c) thresholded markers (white "m" printing); (d) selected features (Color Figure 4.26b); see color insert following page 150.

In order to select only the candies that have the printed "m" on the visible face, a separate marker image was obtained by thresholding the image for the printed white characters. This was done by creating a background with a grey-scale closing to remove the white printing, and then subtracting that background from the original. After thresholding, the markers (Figure 4.26c) can be used in a feature-AND with the binary image of the candies (Figure 4.26b). The result is then used as a mask to recover the candy image, by combining it with the original and keeping whichever pixel is brighter. The black pixels in the selected features are replaced by the color information, and the white background erases everything else.

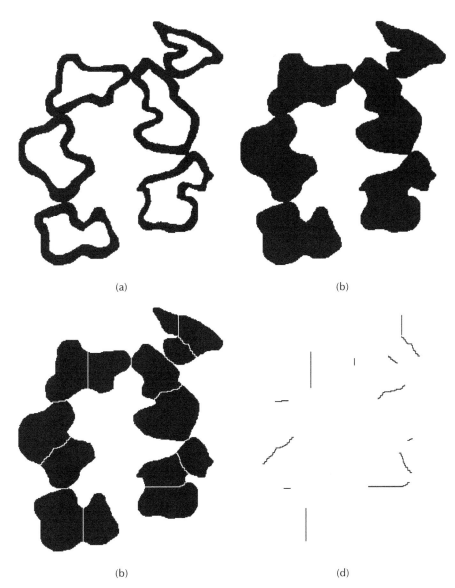

(a)   (b)

(b)   (d)

FIGURE 4.27 Separation of touching, irregular, hollow features using Boolean logic and a watershed segmentation, as described in the text.

COMBINED BOOLEAN OPERATIONS

Boolean operations are often used in combination with each other, or with the various morphological operations described above. All of the Boolean operations, both pixel- and feature-based, are combined in the next example, to solve one of the problems that can arise when attempting to use the watershed segmentation procedure to separate touching features. The image in Figure 4.27 is a thresholded

(e)   (f)

(g)   (h)

FIGURE 4.27 (continued)

binary image of the sheaths around nerve axons. The problem is that the sheaths touch each other, but because they are both hollow and irregular in shape they cannot be straightforwardly separated by the watershed algorithm. The procedure, illustrated in the figure, requires several duplicates of the original image (Figure 4.27a). The first duplicate has internal holes filled (Figure 4.27b). A duplicate of this image is then processed by watershed segmentation (Figure 4.27c), but it can be seen that in addition to the correct lines of separation there are additional lines that cut through the features. Combining images Figure 4.27b and Figure 4.27c using a Boolean exclusive-OR isolates just the separation lines (Figure 4.27d).

A separate image of the holes inside the original features (Figure 4.27e) is prepared by ex-ORing the original (Figure 4.27a) with the filled image (Figure 4.27b). Using these holes as a marker image, a feature-AND with the lines in image (Figure 4.27d) selects just those lines that cross the interior of the features (Figure 4.27f). Restoring these lines to the watershed result (Figure 4.27c) using a Boolean OR corrects for the additional segmentations due to shape (Figure 4.27g). Finally, combining this image with the original using a Boolean AND removes the interior, leaving the hollow, irregular features properly separated (4.27h). This sequence has multiple steps and may appear confusing initially, but serves both as a very useful guide and example to the various Boolean procedures and an illustration of the power that they offer, particularly when used in conjunction with other processing operations.

REGION OUTLINES AS SELECTION CRITERIA

The markers used for a feature-based AND operation shown above were features present in the original image, in one way or another. It is also very useful to generate various kinds of lines to use as a marker for selection of features. One of the most important kinds of lines is the boundary line that separates one region from another. These boundaries in images of cut sections through three-dimensional structures represent surfaces in three-dimensions. Much of the chemistry and many of the mechanical properties of structures depend on those surfaces, and the presence (or absence) or objects from them is often of interest. The measurement of adjacency between different objects is something that can be measured using a combination of morphological and Boolean operations.

Outlines and boundaries are important markers to measure adjacency (common boundaries between features) or to select adjacent features. These are two somewhat different ideas, best illustrated by an example. To emphasize the importance of the difference between the conventional pixel-based Boolean AND and the feature-based AND, Figure 4.28 illustrates two different types of measurement and selection that can be performed. The original image is milk. Casein micelles are positioned around the periphery of the fat globules.

Measurement of the fraction of the surface area of the fat globules occupied by casein particles can be performed by first creating an outline of the pixels just around the globules. This is done by duplicating the binary thresholded image showing just the fat, dilating it, and Ex-ORing the dilated version with the original binary. The dilation is necessary because there are no common pixels between the globules and the particles. The dilation distance must be a minimum of one pixel, but can be made greater to define the distance that is considered to be adjacent or touching. Dilation by thresholding the EDM of the background is preferred over iterative neighbor selection because it is isotropic. The total length of the outlines around the globules, ratioed to the area of picture, is proportional to the surface area of the globules per unit volume as discussed in Chapter 1.

Performing a conventional pixel-based AND of the outlines with the binary image of the dark casein particles selects just those pixels on the outline that are covered by the particles. Measuring the total length of those lines gives the length of the touching interface between the casein and the float globules. In Figure 4.28(e) the portions of the line that touch particles, as selected by an AND, are superimposed

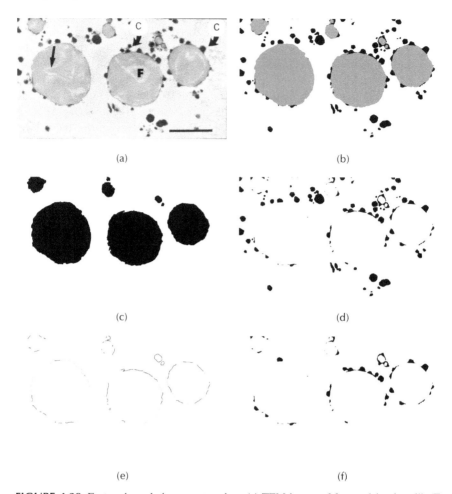

(a)   (b)

(c)   (d)

(e)   (f)

FIGURE 4.28 Feature boundaries as test probes: (a) TEM image of fat emulsion in milk (F = fat globules, C = casein micelles, arrow = fat crystal within the globule, original picture from Goff et al. appears as Figure 7-3a in Stanley and Aguilera, used with permission); (b) thresholded and superimposed fat globules (grey) and micelles (black); (c) fat globules; (d) micelles; (e) layer of pixels around the fat globules obtained by dilating and ex-OR with original; with the superimposed result of ANDing the outlines with the micelles (image d); (f) using the outline of the fat globules as a marker and performing a Feature-AND with image (d).

on the complete outline. Measurement of the length of the touching segments divided by the total length of the outline gives a direct measurement of the fraction of the surface of the fat globules occupied by casein micelles (41.7% in the example). This is usually described as a measurement of adjacency.

Instead of the pixel-based AND, it is also possible to perform a feature-based AND to obtain a very different, but also very useful result. Using the outline as a marker, and combining it with the image of the dark casein particles using a feature-AND, provides a way to select all of the casein particles that touch the fat globules as shown in Figure 4.28(f). The image of the selected particles can then be used for counting, measurement or other analysis as required.

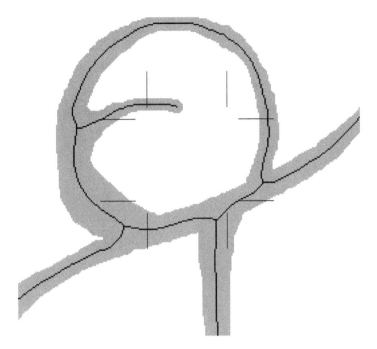

FIGURE 4.29 Illustration of skeletonization.

SKELETONIZATION

A specialized form of erosion can be used to remove pixels from the outside of a thresholded, binary feature leaving just the midline or skeleton. These lines are useful in a variety of situations, ranging from cleaning up tessellations such as that formed by cell walls in plant tissue, to extracting the basic topological shape of features. Figure 4.29 shows a typical skeletonization result: the original feature (shown in grey) is reduced to a line of pixels. The process can be performed with different algorithms that produce slightly different results, but with the same overall effect. An iterative erosion that removes pixels adjacent to background, but only if removing the pixel does not result in breaking the remaining feature into segments, is the most common approach, but somewhat smoother shapes result for the skeleton when it is taken from the ridge lines of the Euclidean distance map.

First, it is useful to consider the role of skeletonization in cleaning up thresholded images. In many situations, the preparation of a sample involves staining, polishing, sectioning of a finite thickness slice, or other techniques that produce cell boundaries which appear much thicker than they actually are. Skeletonization can thin these down to idealized lines, which often simplifies measurement of the structure. In the example of Figure 4.30, the staining of a cross section of pork muscle shows broad endomysium separations between muscles, and also dark interiors of the muscles.

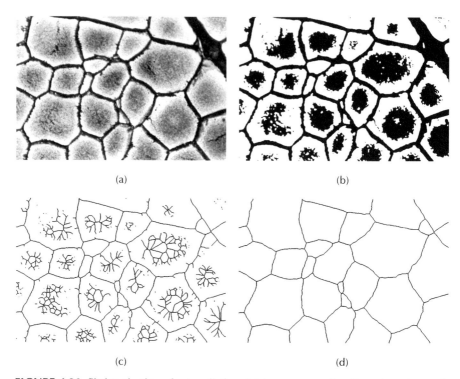

(a)   (b)

(c)   (d)

FIGURE 4.30 Skeletonization of a tessellation: (a) transverse section through pork muscle stained with silver to show endomysium (image courtesy of Howard Swatland, Department of Animal and Poultry Science, University of Guelph); (b) thresholded binary; (c) skeletonized; (d) skeleton pruned to remove all terminal branches and leave just the continuous tessellation.

Thresholding this image (and applying a closing to clean up some noise) does not produce a clean structural image suitable for measurement. Skeletonizing the binary image thins down the tessellation but leaves additional lines in the interior of each region.

These extraneous lines can be removed by pruning the skeleton. To understand this, consider the diagram in Figure 4.29. Most of the pixels in a skeleton have exactly two neighbors out of the eight possible positions. A few may have three or even four; these are usually called nodes or branch points and represent junctions in the network. Pixels that have only a single neighbor are end points. In a tessellation there should be no end points. Pruning a skeleton is a procedure that finds the end points (with one neighbor) and erases them, and then repeats the process until there are none left. As shown in Figure 4.30(d), this removes the isolated lines and branches. In the example, lines that do not have a connection to the continuous network (which touches all edges of the image) have also been eliminated, leaving just the continuous network.

One further note is in order. Skeleton lines a single pixel thick are difficult for printers to show clearly, so in some of these examples the skeleton lines have been dilated to 3 pixels thick so that they show up better in the book. For measurement purposes, they would all be just one pixel wide, and connected as shown in Figure 4.29. The skeleton lines are eight-connected, meaning that they are continuous lines in which a pixel is understood to touch any of its eight possible neighbors. That is also the most common convention for deciding that pixels touch each other to form a connected object. But if an image with an eight-connected skeleton is inverted, so that the cells become features, they are not separated. The pixels in the cells also touch at the corners and cross the skeleton lines. In general, if an eight-connected rule is used for pixels in features, then a four-connected rule (pixels touch only if they share a common side, not just a corner) must be adopted for the background.

In the example of Figure 4.31, a section through potato shows broad cell walls because of the finite thickness of the section and the angles of the walls. In the thresholded image (from the red channel of the original color image), the broad walls and some extraneous dark specks complicate measurement. Skeletonization reduces the walls to lines, and discarding any terminal branches (ones that have end points) and are short (in the example, less than 20 pixels) or lines that are not a part of the continuous tessellation, cleans up the image.

There are still some breaks in the tessellation that may or may not be real, and which can be seen to correspond to gaps in the original image. If the breaks are due to inadequacies of preparation, then completing the tessellation by filling in the gaps is legitimate. Rather than drawing them in by hand, an automatic method that works well if the cells are convex is to invert the image and apply a watershed segmentation to the interiors of the cells. As shown in Figure 4.31(e), this fills in the gaps, but in this case also subdivides a few of the cells that have shapes that are narrow in the center.

Skeletonization of a cell image allows measurement of several characteristics that are often important descriptors of the three-dimensional structure and relate to its macroscopic properties. As an example of the use of the skeleton for a cell structure, Figure 4.32 shows cross-sections (at the same magnification) through apples which are characterized by a sensory test panel as having firm, medium, and soft textures.

In order to correlate structural properties with sensory parameters, the images were thresholded and skeletonized. The principal difference between the firm and medium apples is in the scale of the network structure. The node density in the firm apples is 1.72 per 10,000 μm^2, and this drops to 1.25 for the medium apples. The simple coarsening of the structure allows greater compliance and can account for the change in sensory perception. The structure of the soft apples is quite different. The presence of a few large open spaces allows much more deformation before fracture when the apple is bit into. From a measurement standpoint, the distribution of lengths of the skeleton branches is very different for soft apples. The mean value doubles and the standard deviation quadruples compared to the firm apples, while the statistical kurtosis and appearance of the distribution suggests that a duplex structure with two sizes of structure has formed. There may be other differences between these apple varieties as well (for instance, the area or intercept length of the sections through the cells will show a similar trend), but the ability of simple microstructural measurements to support sensory experience is interesting.

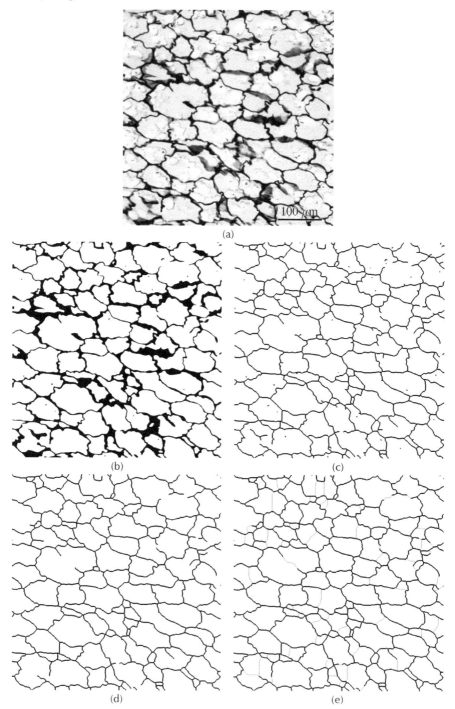

FIGURE 4.31 Editing the skeleton: (a) section of potato (red channel) showing cell walls; (b) thresholded binary; (c) skeletonized; (d) elimination of short terminal branches and disconnected segments; (e) filling in of additional lines by applying a watershed to the inverse image.

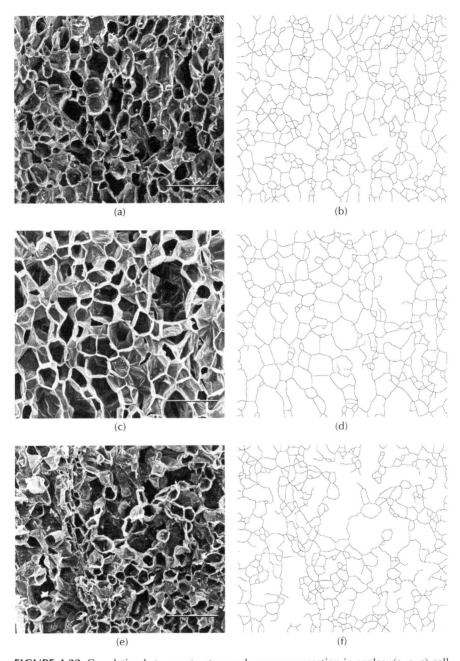

FIGURE 4.32 Correlation between structure and sensory perception in apples: (a, c, e) cell structure in firm, medium and soft apples; (b, d, f) skeletonized structure measured as discussed in text. Magnification marker is 500 µm. (Source: P. Allan-Wojtas, K. Sanford, K. McRae, and S. Carbyn, An Integrated Microstructural and Sensory Approach to Describe Apple Texture, *Journal of the American Horticultural Society,* 128(3), 381–390, 2003. Reproduced with the permission of the Ministry of Agriculture and Agri-Food, Canada.)

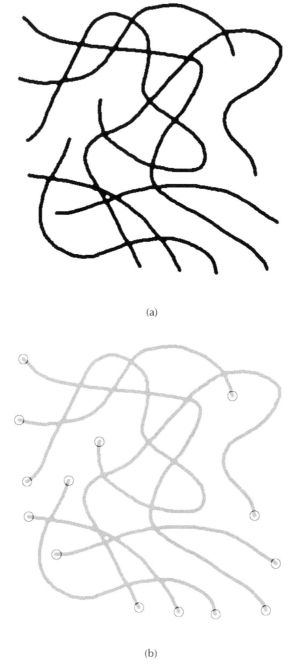

(a)

(b)

FIGURE 4.33 Example of an image with crossing fibers: (a) binary image; (b) end points marked using skeleton (there are 14 end points, corresponding to 7 fibers, and the total length of lines in the skeleton is 21 inches, so the average fiber length is 3 inches).

FIBER IMAGES

Skeletonization is also used with fiber images, because it simplifies counting and measurement. In the example image of Figure 4.33, even a few crossing fibers are visually difficult to count. Counting the end points (with one neighboring pixel) in the skeleton of the fibers gives a direct fiber count because each fiber has two ends. Measuring the total length of the skeleton lines and dividing by the number of fibers (half the number of ends) gives the average fiber length, without any need to disentangle the crossing features. (In the few instances in which measurement of individual crossing fibers is really necessary, the skeleton also plays a role. Finding and removing all of the crossing points — pixels with more than two neighbors — isolates the segments of the fibers so they can be individually measured. Applying logic to match up the segments based on their having end points that lie close together and similar orientations allows adding together the pieces of each fiber to get the overall length. This is a very specialized process, which is only rarely employed.)

In a more complex fiber image such as the network of cellulose fibers in Figure 4.34, the same technique of counting ends still works to determine the number and total length of fibers per unit area. If a fiber has only one end in the image field, it is counted as one-half fiber. That gives the correct number per unit area because the other half would be counted in an adjacent field of view. A few end points may be missed if fiber ends lie exactly on other fibers, but that is primarily a matter of choosing the proper image magnification.

In a branching structure, the individual segments may be of interest (for example to measure their length, orientation, etc.). Skeletonization followed by eliminating the nodes or branch points disconnects all of the segments so that they can be measured as individual features. In the example of Figure 4.35, the root structure of a soybean seedling has been captured by spreading out the roots on a flatbed

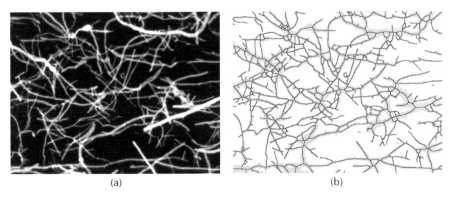

(a)   (b)

FIGURE 4.34 More complex fiber image: (a) original grey scale image of cellulose fibers; (b) skeleton superimposed on binary image (measurement of the skeleton reports 174 end points and 48 inches of total length in an area of 3 square inches).

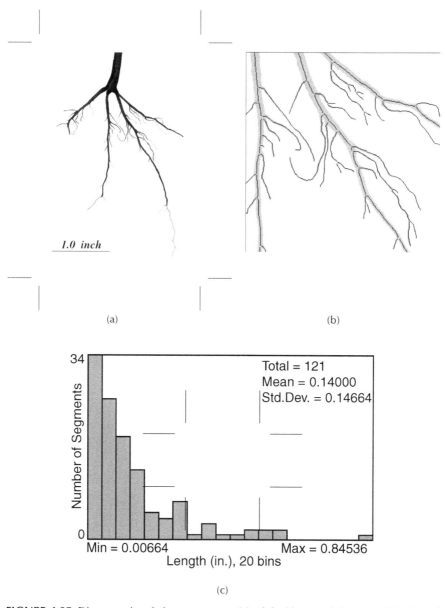

(a)   (b)

(c)

FIGURE 4.35 Disconnecting skeleton segments: (a) original image of plant root; (b) enlarged detail of skeleton with nodes removed, allowing measurement of individual segments; (c) length distribution of segments.

scanner. Thresholding, skeletonization, and removal of the points with more than two neighbors, leaves the individual branches in the root structure. In a healthy plant with adequate moisture and soil nutrients, a distribution with more uniform lengths, rather than a few long and many very short branches, would be expected.

SKELETONS AND FEATURE SHAPE

The skeletons of individual features provide basic topological (shape) information, which will also be used in the next chapter in the discussion of feature measurements. Figure 4.36 shows that the number of end points in the feature skeleton corresponds to the number of points in each star, which is the most recognizable shape characteristic to a human observer.

In a more complex image, the visual recognition of properties such as the number of arms, or the length of the arms, may be more difficult for a person to recognize. But the computer can do so easily using the same properties of the feature skeleton. In the example of Figure 4.37, selection of the features with exactly six arms, or with medium-length external branches (ones that end in an end point as opposed to extending between two nodes in the structure) is illustrated. Combining these two images with an AND selects features that meet both criteria.

FIGURE 4.36 Skeletons of some simple shapes superimposed on the original features. The number of end points in each skeleton measures the basic topological information (number of arms) that is recognized by human vision.

(a)   (b)

(c)   (d)

FIGURE 4.37 Feature selection using the skeleton: (a) hand-drawn star patterns; (b) features with exactly six ends; (c) features with medium length arms; (d) Boolean AND combining (b) and (c).

MEASURING DISTANCES AND LOCATIONS
WITH THE EDM

In the preceding example, another variation present for the various irregular stars is the width of the arms. Measuring the width of an irregular feature is often desired, and sometimes cannot be handled well using a grid. In the example of Figure 4.19 a vertical line grid was appropriate because the thickness of the coating was implicitly defined in that direction. But for the cross-section of the vein in Figure 4.20 the radial lines do not really measure the desired dimension, and for a general, arbitrary shape such as that shown in Figure 4.38, no fixed grid can be devised that would be useful.

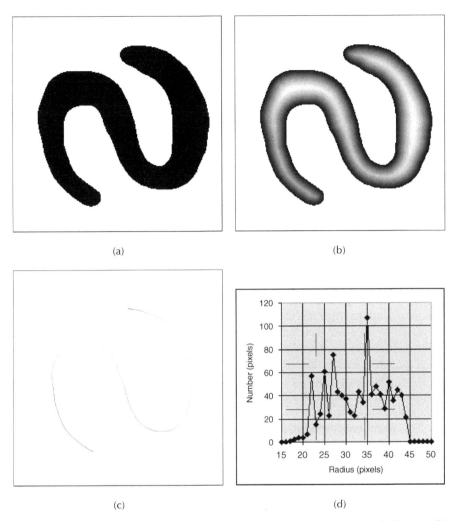

(a)   (b)

(c)   (d)

FIGURE 4.38 Measurement of the width of an irregular feature: (a) example feature; (b) Euclidean distance map; (c) skeleton with EDM values assigned to pixels; (d) histogram of values in the skeleton. The minimum width of the feature is 34 (the lowest non-zero point on the histogram is 17, which is the radius of the inscribed circle) and the maximum is 88 (the highest non-zero point on the histogram is 44). The mean width is 65.5 ± 13.1 (twice the mean radius of 32.73 ± 6.55).

The EDM provides a way to measure the width of such irregular shapes. The distance map values increase from each edge and reach a maximum at the center of the feature, which appears as a ridge line if the EDM is rendered as a surface. The value of the EDM at each point along this line is the radius of an inscribed circle at that location, and so measures the half-width of the feature at that point. Collecting all of the EDM values along the center line thus permits the measurement of the mean, variation (e.g., standard deviation), minimum and maximum widths of the feature.

FIGURE 4.39 The illustration from Figure 4.37(a) with features shaded according to the width of the lines.

The skeleton marks the midline of the feature, so it can be used to collect the desired EDM values, as shown in the figure. One way to accomplish this is to duplicate the image, calculate the EDM of one copy and the skeleton of the second, and then combine them arithmetically keeping whichever pixel value is lighter. All pixels other than the skeleton are erased to white (background) while the EDM values replace the black of the skeleton pixels. A histogram of the image (ignoring all of the white background pixels) can be analyzed to perform the measurement.

Using the same procedure on the image from Figure 4.37, the features can be grey-scale labeled to indicate the mean width of their arms as shown in Figure 4.39. This adds another criterion by which they may be distinguished and sorted.

This measurement technique is particularly useful for characterizing the width of irregular structures seen in cross section as shown in Figure 4.40. It must be remembered that it is a two-dimensional measurement, and does not replace the use of random line intersections as a tool for measurement of the true three-dimensional thickness of layers sectioned at random orientations.

The location of features within cells, organs, or any other space, can also be measured using the EDM. For a regular geometric shape such as a square or circle, calculation of the distance of a feature from the boundary can be performed using analytical geometry. For a typical irregular shape, this method cannot be used, but the EDM can. Assigning the distance map value to each pixel within the cell or region provides a measurement tool for any shape. Combining the EDM values with an image of the centroid location of each feature, as shown in Figure 4.41, allows the distance of each feature from the boundary to be measured using the grey scale value (which can, of course, be calibrated in distance units rather than density).

In the example, the protein and starch granules were thresholded in one copy of the image and the centroid feature of each marked (the ultimate eroded points could also be used; in this example with convex features the difference is unimportant, but for irregular shapes the centroid may not lie within the feature, and the ultimate eroded point may be preferred). A second copy of the image was thresholded for the cell walls, which were then skeletonized and the image inverted so that the

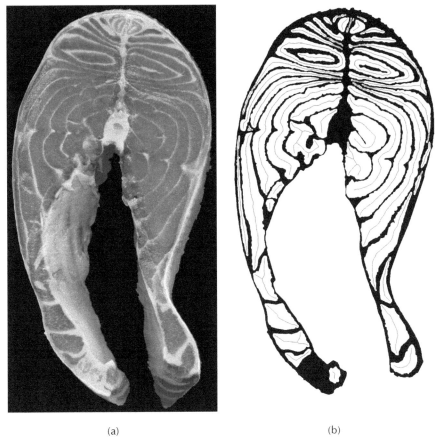

(a)   (b)

FIGURE 4.40 Measurement of the thickness of muscles in farm-raised salmon: (a) scanned cross-section image; (b) binary image after thresholding and opening, with skeleton lines (dilated for visibility) superimposed on the muscles; (c) histogram of the EDM values sampled by the skeleton, which measures the thickness distribution of the muscles.

EDM could be calculated for the interior of the cells. The images were then combined by a mathematical operation that keeps whichever pixel value is brighter in the two images.

Measuring the grey scale value of each feature allows plotting the number of features as a function of distance from the boundary. As shown in Figure 4.41(e), an initial glance at the plot suggests that the features may be preferentially located near cell walls, but this is not necessarily so. It is actually the number per unit area that is of interest, and the size and shape of the cells affects the area as a function of distance. The histogram of the EDM image provides this data as shown in Figure 4.41(f); note that it has the same shape as the plot of number of features. Since it is the ratio of the two plots that measures the tendency of features to locate close to or away from the boundary, in this example the features are uniformly distributed within the cells.

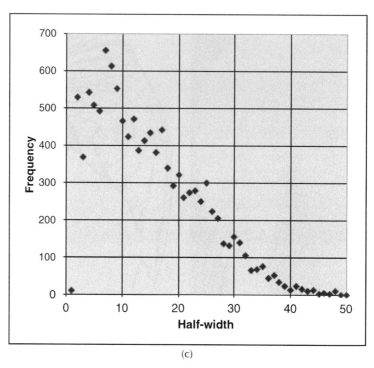

(c)

FIGURE 4.40 (continued)

It is also possible to measure the distance of features from each other. In many instances features are distributed in non-random ways. Clustering of features produces groups separated by gaps, while a self-avoiding distribution has features that are more-or-less uniformly spaced. The key to measuring these tendencies is the random distribution, as will be discussed in detail in Chapter 5. The mean value of the nearest neighbor distance for a random distribution can be calculated as

$$Mean\ Nearest\ Neighbor\ Distance = \frac{0.5}{\sqrt{\dfrac{Number}{Area}}} \tag{4.1}$$

If a feature distribution is clustered, the mean nearest neighbor distance is lower than that for a random distribution, and, vice versa, it is greater for a self-avoiding distribution. For the example image, a plot of nearest neighbor distances is shown in Figure 4.41(g). The mean value of the distribution (4.35 μm) is significantly greater than the calculated value expected for a random distribution (3.20 μm) of the same number of features in the same area, indicating that the features are self-avoiding. More examples of random, clustered, and self-avoiding distributions are shown in Chapter 5. A plot showing all of the nearest neighbor distance and direction vectors (Figure 4.41h) indicates that the spatial distribution is isotropic.

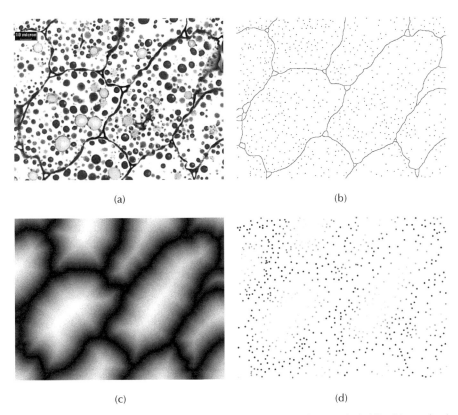

(a)   (b)

(c)   (d)

FIGURE 4.41 Characterizing spatial distributions: (a) original image of toluidine blue stained peanut (red channel from a color image, courtesy of David Pechak, Kraft Foods Technology Center) showing stained protein bodies and starch granules (see Color Figure 1.12; see color insert following page 150); (b) skeletonized cell walls and feature centroids; (c) EDM of cell interior; (d) feature centroids with grey scale values assigned from the EDM; (e) plot of number of features as a function of distance from cell wall; (f) plot of the area as a function of distance; (g) plot of number of features as a function of nearest neighbor distance; (h) rose plot of nearest neighbor distance and direction vectors showing isotropic distribution.

A very similar procedure makes it possible to measure the fractal dimension of clusters. The characteristic of agglomerated clusters is that a plot of the number of features (or the mass of agglomerated particles) as a function of their radial distance from any point in the cluster, plotted on log-log axes, is a straight line whose slope gives the cluster dimension. To measure this, combine the centroids or ultimate eroded points of features such as the crystals in the image of shortening in Figure 4.42 with the EDM of the entire image field. The latter is generated by selecting any arbitrary point (but usually for convenience the center of the image) as a single background point and then calculating the EDM for all other points. Assigning these values to the ultimate eroded points and plotting the number of points as a function of their grey scale value produces the required graph.

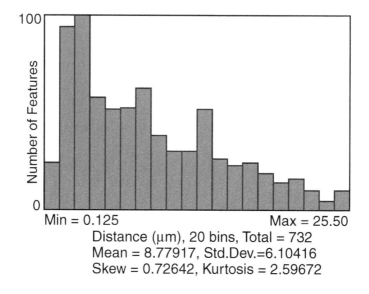

Min = 0.125   Max = 25.50
Distance (μm), 20 bins, Total = 732
Mean = 8.77917, Std.Dev.=6.10416
Skew = 0.72642, Kurtosis = 2.59672

(e)

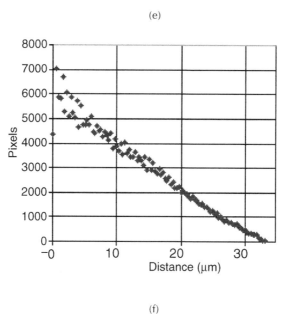

(f)

FIGURE 4.41 (continued)

The number of ways that the EDM, centroids or ultimate eroded points, outlines (and dilated outlines), skeleton, and various points such as end points or nodes, can be combined to perform subtle but important measurements on images is limited only by the imagination. As one more example, consider the illustration in Figure

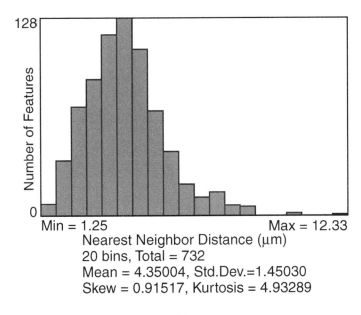

Min = 1.25   Max = 12.33
Nearest Neighbor Distance (μm)
20 bins, Total = 732
Mean = 4.35004, Std.Dev.=1.45030
Skew = 0.91517, Kurtosis = 4.93289

(g)

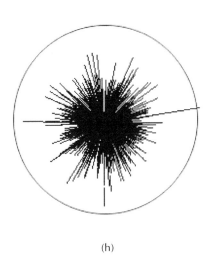

(h)

FIGURE 4.41 (continued)

4.43 of a cell with dendritic processes. Visually, there is an appearance that the length of the terminal branches (ones that have an end) is shorter for those branches that are more distal from the cell body. How can that be verified by measurement?

In this case the skeleton segments can be disconnected by removing the nodes, but only the terminal branches are wanted. So another copy of the skeleton is processed to keep just the end points, which are then used in a feature-based AND

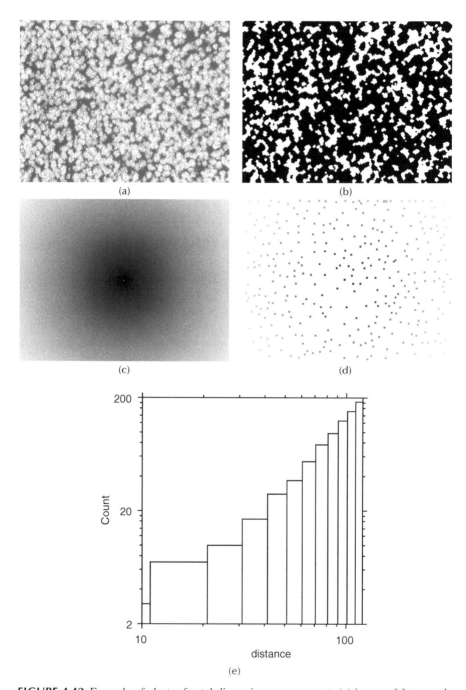

FIGURE 4.42 Example of cluster fractal dimension measurement: (a) image of fat crystals in shortening (courtesy of Diana Kittleson, General Mills); (b) thresholded image (after closing); (c) EDM of the image field; (d) ultimate eroded points with grey scale assigned from the EDM; (e) log-log plot of cumulative count vs. radial distance.

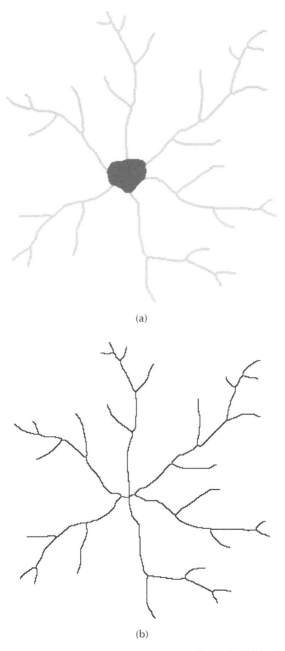

(a)

(b)

FIGURE 4.43 Measurement using skeleton, feature-AND, and EDM: (a) original image illustrating a cell with dendritic extensions; (b) skeleton of the structure; (c) terminal branches produced as described in the text; (d) EDM of the pixels outside the main cell body measuring distance from the cell; (e) grey scale values from the EDM assigned to pixels in the terminal branch skeletons (thickened for visibility); (f) correlation plot showing variation of branch length with distance of the near end of each branch from the cell body.

(c)

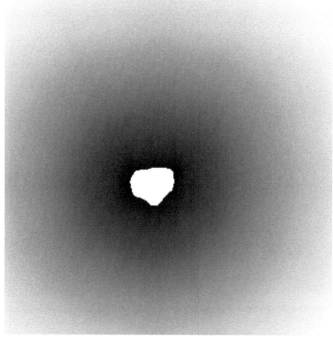

(d)

FIGURE 4.43 (continued)

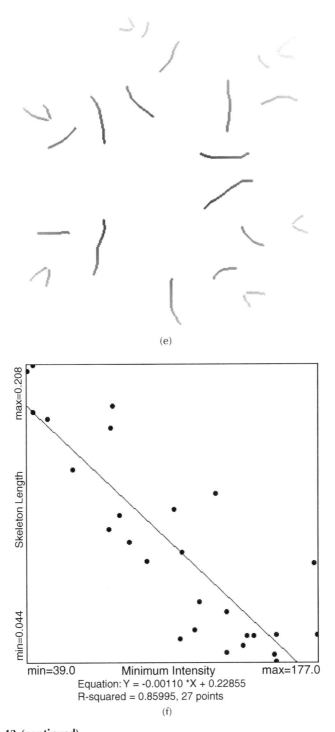

(e)

(f)

FIGURE 4.43 (continued)

to select only those branches that are marked by an end point. These are then combined with the EDM calculated for all pixels outside the cell body. That assigns distance values to the pixels in the skeleton segments that mark the terminal branches. But what should be measured? The most reasonable choice is to plot the length of each terminal segment of the skeleton (calculated as a smooth curve, as discussed in the next chapter) versus the smallest value of grey scale for any pixel in the skeleton. That corresponds to the distance of the nearest point from the cell body. As shown in the figure, the plot indicates a significant correlation between the two variables, confirming the visual impression.

SUMMARY

Morphological and Boolean procedures applied to binary images obtained by thresholding serve multiple purposes. One is simply to clean up imperfect representations of the features or structure present, by removing extraneous noise, smoothing boundaries, filling holes within features, separating touching features, and so forth. Another is to extract skeletons or outlines from images either to use as characterizations of shape and location or to use in conjunction with other images and operations for measurement. A third very important class of operations uses Boolean logic to apply various grids to images so that stereological measurements can be obtained. Many of the global structural measurements (as opposed to the measurement of individual features present in the image, which is covered in the next chapter) that characterize foods (and other materials) can be very efficiently obtained by using these various tools in the appropriate combinations. This chapter has illustrated some of those combinations, but the inventive researcher will find many more useful ones.

5 Measuring Features

Stereological techniques, introduced in Chapter 1, provide effective and efficient measurements of three-dimensional structures from images of sections through them. Many of the procedures can be implemented using computer-generated grids combined with binary images using Boolean logic, as shown in the preceding chapter. Interpretation of the results is straightforward but assumes that proper sampling procedures (e.g., isotropic, uniform and random probes) have been followed.

Another important task for computer-based image analysis is counting, measuring and identifying the features present in images. This applies both to microscope images of sections through structures, and to other classes of microscopic or macroscopic specimens such as particles dispersed on a substrate. It is important to know if the features seen in the images are projections or sections through the objects. Projections show the external dimensions of the objects, and are usually easier to interpret. Some section measurements can be used to calculate size information about the three-dimensional objects as described in Chapter 1, subject to assumptions about object shape.

COUNTING

For both section and projection images, counting of the features present is a frequently performed task. This may be used to count stereological events (intersections of an appropriate grid with the microstructure), or to count objects of finite size. This seems to be a straightforward operation, simply requiring the computer program to identify groups of pixels that touch each other (referred to as "blobs" in some computer-science texts). The definition of touching is usually that pixels that share either a side or corner with any of their eight neighbors are connected. As noted in the previous chapter, when this eight-connected convention is used for the feature pixels, a four-connected relationship is implied for the background, in which pixels are connected only to the four that share an edge.

For images in which all of the features of interest are contained entirely within the field of view, counting presents no special problems. If the result is to be reported as number per unit area, the area of the image covered by the specimen may need to be measured. But in many cases the image captures only a portion of the sample. Hopefully it will be one that is representative in terms of the features present, or enough fields of view will be imaged and the data combined to produce a fair sampling of the objects. When the features extend beyond the area captured in the image, there is a certain probability that some of them will intersect the edges of the image. That requires a special procedure to properly count them.

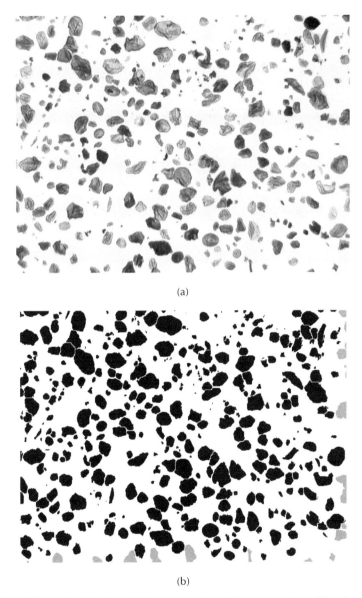

(a)

(b)

FIGURE 5.1 Example of unbiased feature counting: (a) cornstarch particles (courtesy of Diana Kittleson, General Mills); (b) binary image after thresholding and watershed segmentation. The light grey features are not counted because they intersect the bottom or right edges of the image.

The simplest implementation of an unbiased counting procedure is to ignore features that intersect two adjacent edges of the image (for instance, the right side and the bottom) and to count features that reside entirely within the field of view, or are cut off by the other two edges. That produces an unbiased result because the features that cross the right edge of this field of view (and are not counted) would cross the left edge of the next adjacent field, and would be counted, and similarly for the ones at the bottom. So the net count of features per unit area of the image is unbiased and can be used as a representative estimate of the number of features per unit area for the entire sample. Figure 5.1 illustrates this procedure. It is equivalent (in a statistical sense) to counting features that cross any of the four edges of the image as one-half.

If features are so irregular in shape (for instance, long fibers) that they can cross one of the edges where they should be counted, but can loop around and re-enter the field of view across one of the edges where they should not be counted, this method won't work correctly. In that case it is necessary to either find a way (usually a lower magnification image) to identify the features that touch the do-not-count edges and avoid counting them where they reappear, or to find some other unique way to count features (such as the end points of fibers, discussed in the preceding chapter).

There are some situations in which other edge-correction methods are needed, but they are all based on the same logic. For example, in counting the number of pepperoni slices on one slice of pizza, features crossing one cut edge of the slice would be counted and features crossing the other cut edge would not.

As discussed below, a more elaborate correction for edge effects is needed when features are measured instead of just counted.

Measurement typically provides numerical information on the size, shape, location and color or brightness of features. Most often, as noted above, this is done for features that are seen in a projected view, meaning that the outside and outer dimensions are visible. That is the case for the cornstarch particles in Figure 5.1, which are dispersed on a slide. It is also true for the starch granules in Figure 5.2. The surface on which the particles are embedded is irregular and tilted, and also probably not representative because it is a fracture surface. The number per unit area of particles cannot be determined from this image, but by assuming a regular shape for the particles and measuring the diameter or curvature of the exposed portions, a useful estimate of their size can be obtained. This is probably best done interactively, by marking points of lines on the image, rather than attempting to use automatic methods.

The image in Figure 5.2 illustrates the limited possibilities of obtaining measurement information from images other than the ideal case, in which well dispersed objects are viewed normally. Examining surfaces produced by fracture presents several difficulties. The SEM, which is most conveniently used for examining rough surfaces, produces image contrast that is related more to surface slope than local composition and does not easily threshold to delineate the features of interest. Also, dimensions are distorted both locally and globally by the uneven surface topography (and some details may be hidden) so that measurements are difficult or impossible to make. In addition, the surface cannot in general be used for stereological measurements because it hasn't a simple geometrical shape. Planes are often used as ideal sectioning probes into structures, but other regular surfaces such as cylinders can also be used, although they may be more difficult to image. Finally, the fracture surface is probably not representative of the material in general, which is why the fracture followed a particular path.

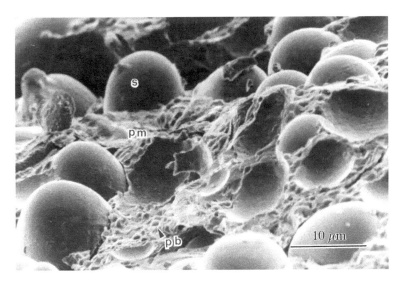

FIGURE 5.2 SEM micrograph of the fracture surface of a bean cotyledon. Starch granules (S) are embedded in a protein matrix (pm). (Original image is Figure 3-23 in Aguilera and Stanley, used with permission.)

These considerations apply to counting and measuring holes as well as particles. Figure 5.3 shows another example of an SEM image of a fractured surface. Note that in addition to the difficulties of delineating the holes in order to measure their area fraction, that would not be a correct procedure to determine the volume fraction using the stereological relationships from Chapter 1. A fracture surface is not a random plane section, and will in general pass through more pores than a random cut would intersect. Also, in this type of structure the large more-or-less round pores are not the only porosity present. The structure consists of small sugar crystals that adhere together but do not make a completely dense matrix, so there is additional porosity present in the form of a tortuous interconnected network. The volume fraction of each of these types of pores can be determined explicitly by measuring the area fraction or using a point grid and counting the point fraction on a planar section. However, from the SEM image it is possible to determine the average size of the sugar crystals by autocorrelation as approximately 5.25 µm, using the same procedure as shown in Chapter 3, Figure 3.43.

In Figure 5.4, the view of the flat surface on which the particles reside is perpendicular, but the density of particles is too high. Not only do many of them touch (and with such irregular shapes, watershed segmentation is not a reliable solution), but they overlap and many of the smaller particles hide behind the larger ones. Dispersal of the particles over a wider area is the surest way to solve this problem, but even then small particles may be systematically undercounted because of electrostatic attraction to the larger ones.

For viewing particles through a transparent medium in which they are randomly distributed, there is a correction that can be made for this tendency of small features to hide behind large ones. Figure 5.5 illustrates the logic involved. The largest

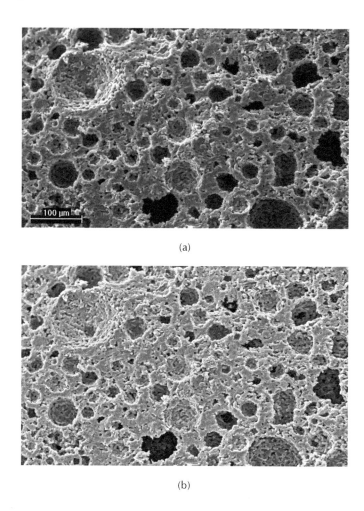

(a)

(b)

FIGURE 5.3 Scanning electron microscope image of the fractured surface of a dinner mint (a). Retinex-based contrast compression (b) described in Chapter 3 makes it possible to see the details inside the dark holes while retaining contrast in the bright areas. (Original image courtesy of Greg Ziegler, Penn State University Department of Food Science.)

features can all be seen, so they are counted and reported as number per unit volume, where the volume is the area of the image times the thickness of the sample. Then the next particles in the next smaller class size are counted, and also reported as number per unit volume. But for these particles, the volume examined is reduced by the region hidden by the larger particles. For the case of a silhouette image, this is just the area of the image minus the area of the large particles, times the thickness. If smaller particles can be distinguished where they lie on top of the large ones, but not behind them, then the correction is halved (because on the average the large particles are halfway through the thickness). The same procedure is then applied to each successive smaller size class. This method does not apply to most dispersals of objects on surfaces because the distribution is usually not random, and the small

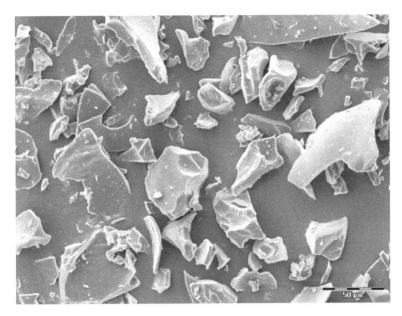

FIGURE 5.4 SEM image of pregelatinized cornstarch, showing small particles that are hidden by or adhering to larger ones.

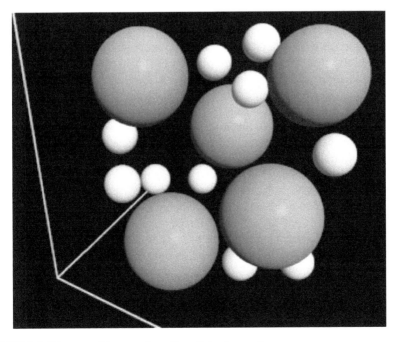

FIGURE 5.5 Diagram of large and small particles in a volume.

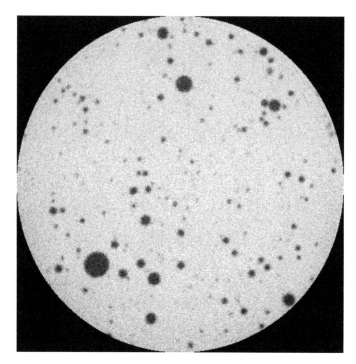

FIGURE 5.6 Tomographic cross-section of an aerated confection (courtesy of Greg Ziegler, Penn State University Department of Food Science). The full width of the image is 14 mm, with a pixel dimension of 27.34 µm.

objects are much more likely to find a hiding place under the large ones. There are a few specialized sample preparation procedures that avoid this problem, in which case the same correction procedure can be used, but usually the problem is solved by dispersing the particles sufficiently to prevent overlaps.

Counting the number of features in an opaque volume requires the use of the disector logic described in Chapter 1. The counting procedure can, of course, be performed manually, but image processing routines can simplify the process as shown in the following example. Figure 5.6 shows one cross section through an aerated confection, obtained by X-ray tomography. In general such images show density differences, and in this case the bubbles or voids are dark. A series of parallel sections can be readily obtained by this approach, just as similar tomography (but at a larger scale) is applied in medical imaging. From a series of such sections, a full three-dimensional rendering of the internal structure can be produced, just as from confocal microscope images or serial sections viewed in the conventional light or electron microscope (except of course for the potential difficulties of aligning the various images and correcting for distortions introduced in the cutting process).

Many different viewing modes, including surface rendering, partial transparency, stereo viewing, and of course the use of color, are typically provided for visualization of three-dimensional data sets, along with rotation of the point of view. Still images such as those in Figure 5.7 do not fully represent the capabilities of these interactive

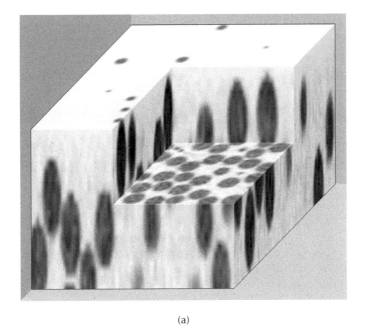

(a)

(b)

FIGURE 5.7 Examples of visualization of a small fragment of a three-dimensional data set produced from sequential tomographic sections (note that the vertical dimension has been expanded by a factor of 4 in these examples): (a) arbitrary surfaces through the data block; (b) sequential sections with the matrix transparent, so that just the holes show; (c) surface rendering of the holes with the matrix transparent.

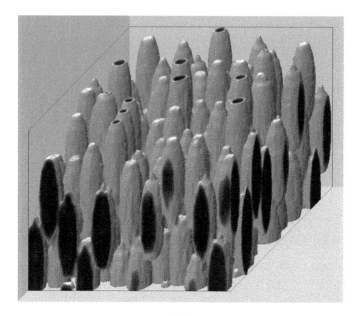

(c)

FIGURE 5.7 (continued)

programs, but it should be emphasized that while the images have great user appeal, and have become more practical with fast computers with large amounts of memory, they are not very efficient ways to obtain quantitative information. Indeed, few of the three-dimensional visualization programs include significant processing or measurement tools.

The disector logic presented in Chapter 1 requires a minimum of two parallel images a known distance apart, such as those in Figure 5.8. Because the individual sections are grainy (noisy) in appearance, processing was applied. The maximum likelihood method introduced in Chapter 3 sharpens the brightness change at the edges of the holes and reduces the grain in the image. making it easier to threshold the holes. The disector count for convex features like these holes is simply the number of intersections that are present in one section but not in both. The Feature-AND logic presented in Chapter 4 identifies the features present in both images (holes that continue through both sections) even if they are different in size or shape. Removing these features that pass through both section leaves those corresponding to holes for which either end lies in the volume between the slices, which are then counted.

There are 65 features (intersections) counted, so the number of holes per unit volume according to Equation 1.9 in Chapter 1 is 65/2 divided by the volume sampled. That volume is just the area of the section image times the distance between the sections, which is 9.7 mm^3 (6.3 μm times 1.54 cm^2). Consequently, N_V is 3.35 holes per cubic millimeter. Since the area fraction of holes in a section can also be measured (6.09%) to determine the total volume fraction of holes, the mean volume of a hole can be determined as 0.0018 mm^3. Of course, that measurement is based

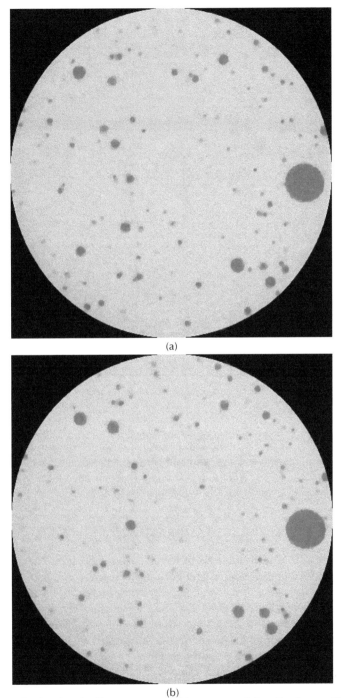

(a)

(b)

FIGURE 5.8 Performing the disector count: (a, b) two sequential section images 63 μm apart, processed with a maximum likelihood operator to sharpen the edges of the holes; (c) after thresholding and Boolean logic, the binary image of the 65 holes that pass through either one of the sections but not through both of them.

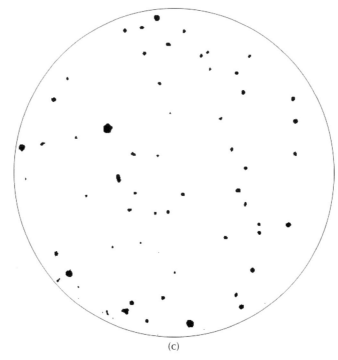

(c)

FIGURE 5.8 (continued)

on using just two of the sections, and the precision can be improved by processing more pairs of planes, so that the total number of events (feature ends) is increased, as discussed in Chapter 1. Additional sampling is also needed if the specimen is not uniform.

CALIBRATION

After the various measurements of feature size, shape, position, and brightness or color described in this chapter have been made, the resulting data are often used for feature identification or recognition. There are several different approaches for performing this step, which will be discussed at the end of the chapter. All of these procedures require image calibration. Both counting results, which are usually reported as number per unit area, and feature measurements (except for some shape descriptors) rely on knowing the image magnification. In principle, this is straightforward. Capturing an image of a known scale, such as a stage micrometer, or a grid of known dimension (many of which are traceable standards to NIST or other international sources) allows easy determination of the number of pixels per micron, or whatever units are being used. For macroscopic images, such as those obtained using a copy stand, rulers along the sides or even a full grid on which objects are placed are commonly employed.

The assumption, of course, is that the images of the objects of interest have the same magnification everywhere (no perspective distortion from a tilted view, or pincushion distortion from a closeup lens, or different magnification in the X and Y directions from using a video camera and framegrabber, or area distortions because of piezo creep in an AFM, for example). If the scale is photographed once and then the calibration is stored for future use, it is assumed that the magnification does not change. For a light microscope, that is usually a good assumption because the optics are made of glass with fixed curvatures and, unless some adjustments are made in the transfer lens that connects the camera to the scope, the same image magnification will be obtained whenever a chosen objective lens is clicked into place. Focusing is done by moving the sample toward or away from the lens, not altering the optics.

For a camera on a copy stand, it is usually possible to focus the camera lens, which does change the magnification. But on the copy stand the scale is usually included in the image so there is always a calibration reference at hand. Electron microscopes use lenses whose magnification is continuously adjustable, by varying the currents in the lens coils. Even in models that seem to provide switch-selected fixed magnifications, the actual magnification varies with focus, and also depends on the stability of the electronics. Keeping the magnification constant over time, as the calibration standard is removed and the sample inserted, is not easy (and not often accomplished).

Calibrating the brightness or color information is much more difficult than calibrating the image magnification. Again, standards are needed. For density measurement, film standards with known optical density are available from camera stores (Figure 5.9), but fewer choices are available for microscopic work. For color comparisons, the Macbeth chart shown in Chapter 2 is a convenient choice, but again is useful only for macroscopic work. A few microscope accessory and supply companies do offer a limited choice of color or density standards on slides. For electron microscopy, only relative comparisons within an image are generally practical. In images such as those produced by the AFM in which pixel value represents elevation, calibration using standard artefacts is possible although far from routine. Some of the other AFM modes (tapping mode, lateral force, etc.) produce signals whose exact physical basis is only partially understood, and cannot be calibrated in any conventional sense.

One difficulty with calibrating brightness in terms of density for transmission images, or concentration for fluorescence images, etc., is that of stability. The light sources, camera response, and digitization of the data are not generally stable over long periods of time, nor can they be expected to repeat after being turned off and on again. As mentioned in Chapter 2, some cameras are not linear in their response and the relationship between intensity and output signal varies with signal strength. As an example, a camera may be linear at low light levels and become increasingly non-linear, approaching a logarithmic response at brighter levels. Some manufacturers do this intentionally to gain a greater dynamic range. If automatic gain and brightness circuitry, or automatic exposure compensation is used, there will be no way to compare one image to another.

The only solution in such cases is to include the standard in every image. With a macro camera on a copy stand, or with a flatbed scanner as may be used for reading films or gels, that is the most reasonable and common way to acquire images. The calibration steps are then performed on each image using the included calibration data. Obviously, this is much more difficult for a microscopy application.

The procedure for establishing a calibration curve is straightforward, if tedious. It is illustrated in Figure 5.9, using an image of a commercial film density wedge

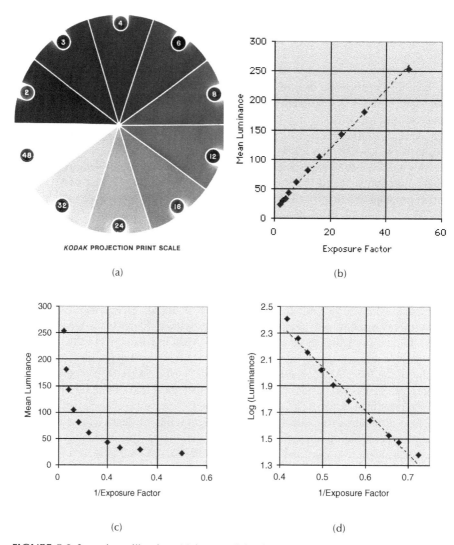

FIGURE 5.9 Intensity calibration: (a) image of density wedge; (b) plot of mean luminance of each region vs. labeled exposure factor; (c) luminance vs. reciprocal of exposure factor (optical density); (d) log (luminance) vs. reciprocal of exposure factor.

used for darkroom exposure settings. This was scanned with a flatbed scanner, and in normal use would have been included in the same scanned image as the film or gel to be quantified. The wedge has ten areas of more-or-less uniform density, with known (labeled) exposure factor settings (the exposure factor is proportional to the inverse of the optical density). Measuring the mean pixel brightness value in each region allows creating a calibration curve as shown in the figure.

Plotting the mean pixel value (luminance) vs. the labeled exposure factor produces a graph that appears quite linear and could certainly be used for calibration, but other plots reveal more of the principal of optical density. The luminance drops exponentially with density as shown in Figure 5.9(c), following Beer's law. Replotting the data with the log of the luminance as a function of optical density produces a nearly linear relationship (ideally linear if the wedge, detector, electronics, etc. are perfect) that spreads the points out more uniformly than the original plot of the raw data. Once the curve is constructed, it can then be stored and used to convert points or features in measured images to optical density.

Other types of standards can be used for other applications. Drug doses or chemical concentration (for fluorescence images), average atomic number (for SEM backscattered electron images), step height (AFM images), etc. can all be quantified if suitable standards can be acquired or fabricated. In practically all cases the calibration curves will be non-linear. That means that the brightness value for each pixel should be converted to the calibrated units and then the average for the feature calculated, rather than the mean pixel value being used to look up a calibrated value. Other statistics such as the standard deviation, minimum and maximum are also useful as will be shown in some of the examples below.

Color is a more challenging problem. As pointed out in Chapter 2, a tristimulus camera that captures light in three relatively broad (and overlapping) red, green, and blue bands cannot be used to actually measure color. Many different input spectra would be integrated across those bands to give the exact same output. Since human vision also uses three types of detectors and performs a somewhat similar integration, it is possible to use tristimulus cameras with proper calibration to match one color to another in the sense that they would appear the same to a human observer, even if their color spectra are not the same in detail. That is accomplished by the tristimulus correction shown in Chapter 2, and is the principal reason for using standard color charts.

In most image analysis situations, the goal is not to measure the exact color but to use color as a means of segmenting the image or of identifying or classifying the features present. In Figure 5.10, the colored candy pieces can be thresholded from the grey background based on saturation. As pointed out in Chapter 3, the background is more uniform than the features and can be selected relatively easily. After thresholding, a watershed segmentation is needed to separate the touching features. Then the binary image is used as a mask to select the pixels from the original (combining the two images while keeping whichever pixel value is brighter erases the background and leaves the colored features unchanged).

Measuring the red, green, and blue values in the acquired color image is not very useful, but converting the data to hue, saturation and intensity allows the identification, classification and counting of the candies. As shown in the plot, the

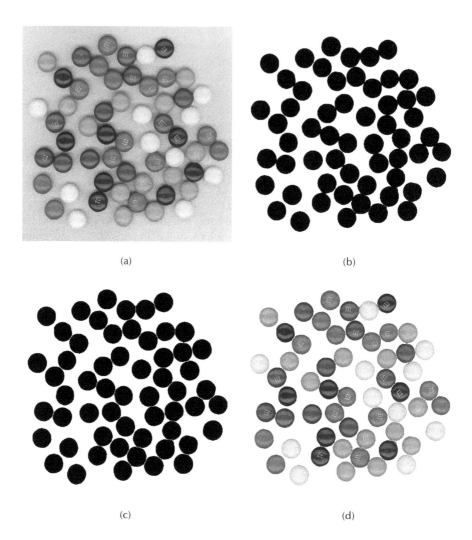

FIGURE 5.10 Measurement of feature color: (a) original image of candies (see Color Figure 3.67; see color insert folllowing page 150); (b) thresholded; (c) watershed segmentation; (d) isolated individual features (see Color Figure 4.26a; see color insert following page 150); (e) plot of hue vs. luminance (brightness) for each feature, showing clusters that count candies in each class.

hue values separate most of the features while the intensity or luminance is needed to distinguish the brown from the orange colors.

There is more information available in this image as well. The ratio of the maximum to minimum luminance for each feature is a measure of the surface gloss of the candy piece. That is true in this instance because the sample illumination consisted of two fluorescent tubes in the flatbed scanner that moved with the detector bar and thus lit each piece of candy with the same geometry. Another way to measure surface gloss is to light the surface using a polarizer and record two images with a

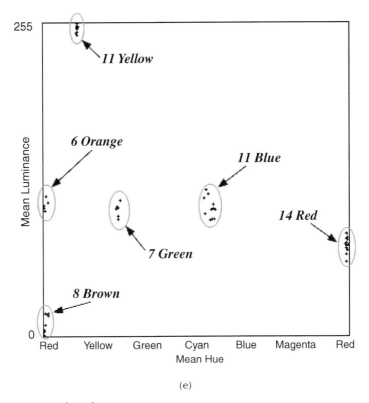

(e)

FIGURE 5.10 (continued)

polarizing filter set to parallel and perpendicular orientations. The ratio gives a measure of surface gloss or reflectivity.

For other classes of particles, the standard deviation of the brightness values within each feature is often a useful measure of the surface roughness.

SIZE MEASUREMENT

Size is a familiar concept, but not something that people are actually very good at estimating visually. Judgments of relative size are strongly affected by shape, color, and orientation. Computer measurement can provide accurate numerical values (provided that good measurement algorithms are used, which is not always the case), but there is still a question about which of many size measurements to use in any particular situation. The area covered by a feature can be determined by counting the number of pixels and applying the appropriate calibration factor to convert to square micrometers, etc. But should the area include any internal voids or holes? Should it include indentations around the periphery? As shown in Figure 5.11, the computer can measure the net area, filled area or convex area but it is the user's problem to decide which of these is meaningful, and that depends on an understanding of the application.

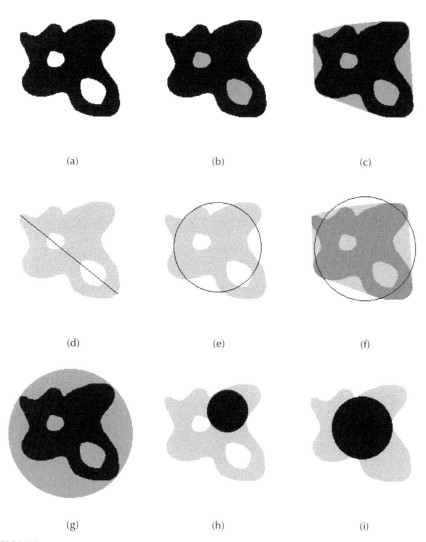

FIGURE 5.11 Measures of size: (a) net area; (b) filled area; (c) convex area; (d) maximum caliper dimension; (e) equivalent circular diameter (net area); (f) equivalent circular diameter (convex area); (g) circumscribed circle; (h) inscribed circle; (i) inscribed circle (filled area).

In the majority of cases, people prefer to work with a size parameter that is linear rather than squared (e.g., mm rather than mm²). One of the commonly used measures is the equivalent circular diameter, which is just the diameter of a circle whose area would be the same as that of the feature (usually the net area, but in principle this could be the filled or convex area as well). But there are several other possibilities. One is the maximum caliper dimension (often called the maximum Feret's diameter) of the feature, which is the distance between the two points that are farthest apart. Another is the diameter of the largest inscribed circle in the feature.

Yet another is the diameter of the smallest circumscribed circle around the feature. All of these can be measured by computer software, but once again it is up to the user to determine which is most suitable for a particular application.

For those interested in the details, the convex area is determined by fitting a taut-string or rubber-band boundary around the feature, as shown in Figure 5.11(c), to fill in indentations around the periphery. It is constructed as a many-sided polygon whose sides connect together the extreme points of the feature on a coordinate system that is rotated in small angular steps (e.g., every 10 degrees). The circumscribed circle (Figure 5.11g) is fit by analytical geometry using the vertices of the same polygon used for the convex hull. The maximum caliper dimension (Figure 5.11d) is also obtained from those vertices. The inscribed circle (Figure 5.11h and 5.11i) is defined by the maximum point in the Euclidean distance map of the feature, but as shown in the illustration, is strongly influenced by the presence of internal holes.

There are some other dimensions that can be measured, such as the perimeter and the minimum caliper dimension. The perimeter can present problems both in measurement and in interpretation. For one thing, many objects of interest are rough-bordered and the length of the perimeter will increase as the magnification is increased. In the particular case in which the increase in length with increasing magnification produces a straight line plot on log-log axes (a Richardson plot), the feature is deemed to have a fractal shape and the actual perimeter length is undefined.

A second difficulty with perimeter measurements is the selection of a measuring algorithm that is accurate. When counting pixels to measure area, it is convenient to think of them as being small squares of finite area. But if that same approach is used for perimeter, the length of the border around an object would become a city-block distance (as shown in Figure 5.12) that overestimates the actual perimeter (in fact, it would report a perimeter equal to the sides of a bounding box for any shape).

Many programs improve the procedure somewhat by using a chain-code perimeter, either one that touches the exterior of the square pixels or one that runs through the centers of the pixels as shown in the figure, summing the links in a chain that are either of length 1 (the side of a square pixel) or length 1.414 (the diagonal of a square pixel). This is still not an accurate measurement of the actual perimeter, produces a length value that varies significantly as a feature it rotated in the image, and also measures the perimeter of a hole as being different from the perimeter of a feature that exactly fills it.

The most accurate perimeter measurement method fits a locally smooth line along the pixels while keeping the area inside the line equal to the area based on the pixel count. This method also rounds corners that would otherwise be perfect 90° angles, and is relatively insensitive to feature or boundary orientation, but it requires more computation than the other methods. The main use of the perimeter value is usually in some of the shape descriptors discussed below.

For a feature such as the irregular S-shaped fiber in Figure 5.13, the length measured as the maximum caliper dimension has little meaning. It is the length along the midline of the curved feature that describes the object. Likewise, the width is not the minimum caliper dimension but the mean value of the dimension perpen-

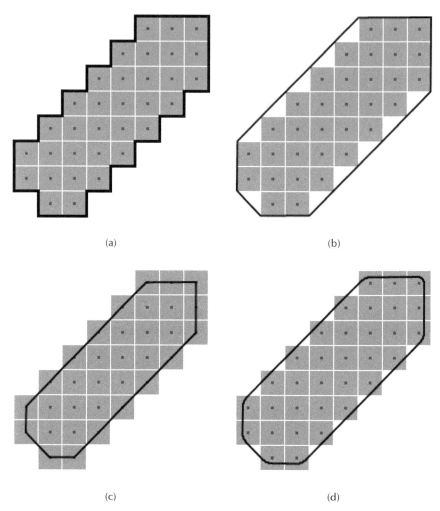

FIGURE 5.12 Perimeter measurement algorithms (shown by dark line) applied to a feature composed of pixels (grey squares with centers indicated): (a) city-block method; (b) chain code surrounding pixels; (c) chain code through pixel centers; (d) area-preserving smooth (super-resolution) method.

dicular to that midline and averaged along it. The skeleton and Euclidean distance map provide these values, as described in the preceding chapter. The length of the skeleton, measured by a smooth curve constructed in the same way as the accurate perimeter measurement shown above, follows the fiber shape. The skeleton stops short of the exact feature end points, actually terminating at the center of an inscribed circle at each end. But the value of the Euclidean distance map at the end pixel is exactly the radius of that circle, and adding back the EDM values at those end points to the length of the skeleton provides an accurate measure of fiber length.

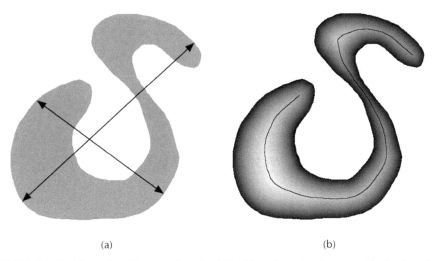

(a)   (b)

FIGURE 5.13 Measuring fiber length and width: (a) an irregular feature with the (largely meaningless) maximum and minimum caliper dimensions marked; (b) the feature skeleton superimposed on the Euclidean distance map. Their combination provides measures for the length and width of the shape.

Similarly, the EDM values along the midline of the feature measure the radii of inscribed circles all along the fiber. The skeleton selects the pixels along the midline, so averaging the EDM values at all of the pixels on the skeleton provides an accurate measurement of the fiber width (and other statistics such as the minimum, maximum and standard deviation of the width can also be determined).

Given this very large number of potential measurements of feature size, which ones should actually be used? Unfortunately, there is no simple guide. Sometimes the definition of the problem will specify the appropriate measurement. The grading of rice as long- or short-grain is defined by the U.S. Department of Agriculture in terms of the length (the maximum caliper dimension) of the grains, so that is the proper measurement to use.

For the example in Figure 5.14, the sample preparation was done by sprinkling rice onto a textured (black velvet) cloth attached to a vibrating table. The vibrations separated the grains and the cloth provided a contrasting background, so the resulting image could be automatically thresholded without any additional processing. Measurement of the lengths of all the grains that were entirely contained in the field of view produced the histogram shown in the figure, which indicates that this is a long grain rice (short grained rice would have a more than 5% shorter than 6 mm) with a more-or-less normal (Gaussian) distribution of lengths.

In this particular case, ignoring the features that intersect the edge of the image does not introduce a serious error (because the range of feature sizes is small, and it is the percentage of short ones that is of greatest interest). However, we will return to the problem of edge-touching features below.

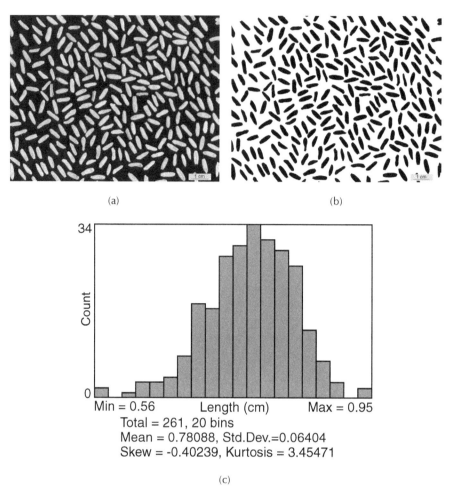

(a)   (b)

(c)

FIGURE 5.14 Measurement of the length of rice grains: (a) original image; (b) thresholded; (c) histogram of measurements.

SIZE DISTRIBUTIONS

Distributions of the sizes for natural objects such as rice grains often produce normal distributions. Statisticians know that the central limit theorem predicts that in any population for which there are a great many independent variables (genetics, nutrients, etc.) that affect a property such as size, the result tends toward a normal distribution. The statistical interpretation of data is somewhat beyond the intended scope of this text, but statistical properties such as the skew (ideally equal to 0) and kurtosis (ideally equal to 3) are often used to judge whether a distribution can be distinguished from a Gaussian, and a chi-squared test may also be used.

Another common type of distribution that arises from many thermodynamic processes is a log-normal curve, with a few large features and many small ones. The ice crystals in ice cream shown in Figure 5.15 provide an example. The sizes (areas

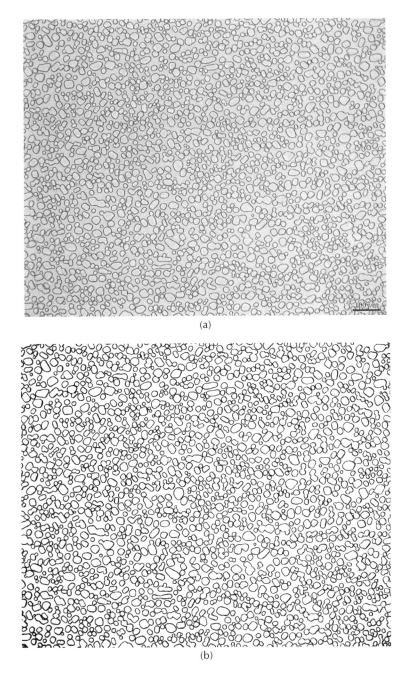

(a)

(b)

FIGURE 5.15 Ice crystals in ice cream (after some remelting has caused rounding of corners); (a) original image, courtesy of Ken Baker, Ken Baker Associates); (b) thresholding the feature outlines; (c) filling the interiors and performing a watershed segmentation; (d) distribution of sizes (area, with measurements in pixels); (e) replotted on a log scale with a superimposed best-fit Gaussian curve for comparison.

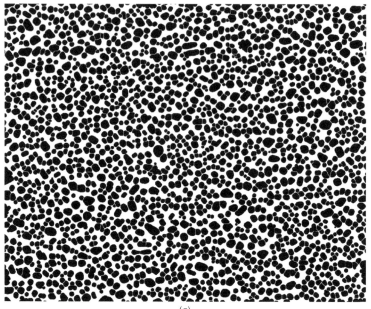

(c)

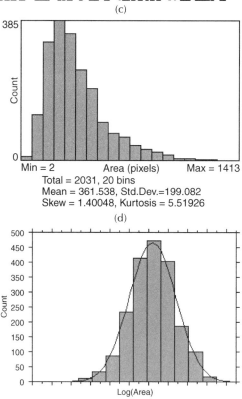

Min = 2   Area (pixels)   Max = 1413
Total = 2031, 20 bins
Mean = 361.538, Std.Dev.=199.082
Skew = 1.40048, Kurtosis = 5.51926

(d)

Log(Area)

(e)

FIGURE 5.15 (continued)

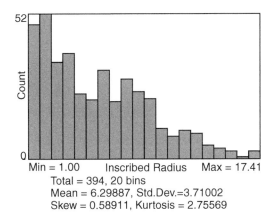

(a)

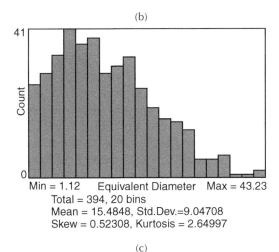

(c)

FIGURE 5.16 Size measurements on the cornstarch particles in Figure 5.1: (a) inscribed circle; (b) circumscribed circle; (c) equivalent circular diameter.

measured in pixels) produce a plot that is skewed toward the right, but plotting instead a histogram of the log values of the feature areas produces a Gaussian result. The image in this example was processed by thresholding of the outlines of the crystals, filling in the interiors, and applying a watershed segmentation.

But not all feature size distributions are Gaussian or log-normal. The cornstarch particles shown in Figure 5.1 were measured using equivalent circular diameter, inscribed circle radius, and circumscribed circle radius. The histograms (Figure 5.16) do not have any easily-described shape, and furthermore each parameter produces a graph with a unique appearance. It is not easy to characterize such measurements, which often result when particle sizes and shapes vary widely.

COMPARISONS

When distributions have a simple shape such as a Gaussian, there are powerful and efficient (and well known) statistical tools that can be used to compare different groups of features to determine whether they come from the same parent population or not. For two groups, the student's t-test uses the means, standard deviations and number of measurements to calculate a probability that the groups are distinguishable. For more than two groups, the same procedure becomes the Anova (analysis of variance). These methods are supported in most statistical packages and can even be implemented in spreadsheets such as Microsoft Excel®.

For measurements that do not have a simple distribution that can be described by a few parameters such as mean and standard deviation, these methods are inappropriate and produce incorrect, even misleading results. Instead, it is necessary to use non-parametric statistical comparisons. There are several such methods suitable for comparing groups of measurements. The Mann-Whitney (also known as Wilcoxon) procedure ranks the individual measurements from two groups in order and calculates the probability that the sequence of groups numbers could have been produced by chance shuffling. It generalizes to the Kruskal-Wallis test for more than two groups. These methods are most efficient for relatively small numbers of observations.

When large numbers of measurements have been obtained, as they usually are when computer-based image analysis methods are employed, the ranking procedure is burdensome. A much more efficient technique for large groups is the Kolmogorov-Smirnov (K-S) test. This simply compares the cumulative distribution plots of the two groups to find the greatest difference between them and compares that value to a test value computed from the number of observations in each group. The probability that the two groups do not come from the same population is then given directly.

Figure 5.17 illustrates the procedure for size data from two different groups of measurements. The measurements of length have been replotted as cumulative graphs, so that for each size the height of the line is the fraction of the group that has equal or smaller values. In this specific case the horizontal axis has been converted to a log scale, which does not affect the test at all since only the vertical difference between the plots matters. The horizontal axis (the measured value) has units, but the vertical axis does not (it simply varies from 0 to 1.0). Consequently, the maximum difference between the two plots, wherever it occurs, is a pure number, which is used to determine the probability that the two populations are different. If

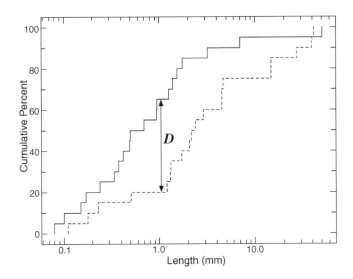

FIGURE 5.17 Principle of the Kolmogorov-Smirnov nonparametric test. The largest differ-ence D that occurs between cumulative plots of the features in the two distributions is used to compute the probability that they represent different populations.

the difference value D is greater than the test value, then at that level of confidence the two groups of measurements are probably from different populations.

Tables of K-S test values for various levels of significance can be found in many statistical texts (e.g., D. J. Sheskin, *Parametric and Nonparametric Statistical Pro-cedures*, Chapman & Hall/CRC, Boca Raton, FL, 2000), but for large populations can be calculated as

$$S = A \cdot \sqrt{\frac{n_1 + n_2}{n_1 \cdot n_2}} \tag{5.1}$$

where n_1 and n_2 are the number of measurements in each group, and A depends on the level of significance. For values of probability of 90, 95 and 99%, respectively, A takes values of 1.07, 1.22 and 1.52.

Other nonparametric tests for comparing populations are available, such as the D'Agostino and Stephens method. The K-S test is widely used because it is easy to calculate and the graphics of the cumulative plots make it easy to understand. The important message here is that many distributions produced by image measurements are not Gaussian and cannot be properly analyzed by the familiar parametric tests. This is especially true for shape parameters, which are discussed below. It is con-sequently very important to use an appropriate non-parametric procedure for statis-tical analysis.

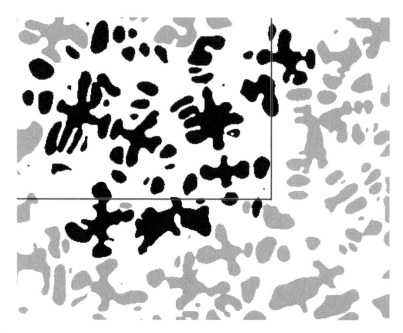

FIGURE 5.18 Use of a guard frame for unbiased measurement. The straight lines mark the limits of the active measurement area. Features that lie in the guard frame region or intersect any edge of the image are neither measured nor counted.

EDGE CORRECTION

The measurements shown in Figures 5.15 and 5.16 were done with a correction for edge-touching features. In the section above on counting, a procedure that counted features intersecting two edges and ignored ones intersecting the other two edges was used. But it is not possible to measure the features that intersect any edge, because there is no information beyond the edge. Since large features are more likely to intersect an edge than small ones, measurement and counting with no edge correction would produce a biased set of measurements that systematically under-represented large features. There are two solutions to this problem.

The older edge-correction procedure is based on the counting correction method. As shown in Figure 5.18, a guard region is drawn inside the image. Features that cross the lines that mark the edge of the active counting region are measured and counted normally, as are all features that reside entirely within the active region. Features that lie inside the guard region are not measured or counted, nor is any feature that touches the edge of the image. Consequently, the guard frame must be wide enough that no feature ever crosses the edge of the active region (which requires that it be measured and counted) and also reaches the edge of the image (which means that it cannot be measured). That means the guard frame must be wider than the maximum dimension of any feature present.

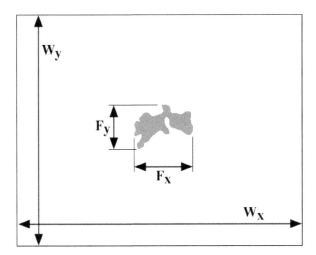

FIGURE 5.19 When every feature within the image is measured, an adjusted count must be used. The dimensions of the image (W_x and W_y) and the projected dimensions of the feature (F_x and F_y) are used to calculate the adjusted count according to Equation 5.2.

Because the guard frame method reduces the size of active counting region compared to the full image area, and results in not measuring many features that are present in the image (because they lie within the guard frame), many computer-based systems that work with digital images use the second method. Every feature that can be measured (does not intersect any edge of the field of view) is measured and counted. But instead of counting in the usual way with integers, each feature is counted using real numbers that compensate for the probability that other similar features, randomly placed in the field of view, would intersect an edge and be unmeasurable.

As shown in Figure 5.19 and Equation 5.2, the adjusted count is calculated from the dimensions of the field of view and the projected dimensions of the feature in the horizontal and vertical directions. For small features, the adjusted count is close to 1. But large features produce adjusted counts that can be significantly greater than 1, because large features are more likely to intersect the edges of the field of view. When distributions of feature measurements are plotted, the adjusted counts are used to add up the number of features in each bin. Statistical calculations such as mean and standard deviation must use the adjusted count as a weight factor for each feature, complicating the math slightly.

$$Adj.\ Count = \frac{W_X \cdot W_Y}{(W_X - F_X) \cdot (W_Y - F_Y)} \tag{5.2}$$

The use of the adjusted count changes the shape of distributions slightly and can markedly alter statistical values. Figure 5.20 shows the results for equivalent circular diameter on the features within the entire image (not using a guard frame) in Figure 5.18. Notice that the total count of features (ones that do not intersect any

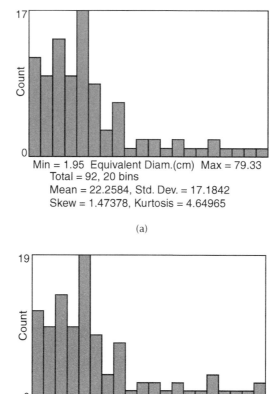

Min = 1.95 Equivalent Diam.(cm) Max = 79.33
Total = 92, 20 bins
Mean = 22.2584, Std. Dev. = 17.1842
Skew = 1.47378, Kurtosis = 4.64965

(a)

Min = 1.95 Equivalent Diam.(cm) Max = 79.33
Total = 106, 20 bins
Mean = 24.8842, Std. Dev. = 19.1968
Skew = 1.24078, Kurtosis = 3.63249

(b)

FIGURE 5.20 Effect of edge correction on the measurement data from the features in Figure 5.18: (a) without edge correction; (b) with edge correction.

edge) is 92 without edge correction, but 106 with the correction, and the mean value also increases by more than 10%.

As pointed out above, large features are more likely than small ones to intersect the edge of the measuring field so that they cannot be measured. Features whose widths are more than 30% of the size of the image will touch an edge more often than not. Note that it is absolutely not permitted to adjust the position of the camera, microscope stage, or objects to prevent features from touching the edge. In general, any fiddling with the field of view to make the measurement process easier, or select particular features based on aesthetic grounds, or for any other reason, is likely to bias the data. There is a real danger in manual selection of fields to be measured. The implementation of a structured randomized selection procedure in which the opportunity for human selection is precluded is strongly encouraged to avoid introducing an unknown but potentially large bias in the results.

Edge intersection of large features can also be reduced by dropping the image magnification so that the features are not as large. But if the image also contains small features, they may become too small to cover enough pixels to provide an accurate measurement. Features with widths smaller than about 20 pixels can be counted, but their measurement has an inherent uncertainty because the way the feature may happen to be positioned on the pixel grid can change the dimension by 1 pixel (a 5% error for something 20 pixels wide). For features of complex shape, even more pixels are needed to record the details with fidelity.

Figure 5.21 shows three images of milk samples. After homogenization, all of the droplets of fat are reduced to a fairly uniform and quite small size, so selection of an appropriate magnification to count them and measure their size variation is straightforward. The ratio of maximum to minimum diameter is less than 5:1. But before homogenization, depending on the length of time the milk is allowed to stand while fat droplets merge and rise toward the top (and depending on where the sample is taken), the fat is present as a mixture of some very large and many very small droplets. In the coarsest sample (Figure 5.21c) the ratio of diameters of the largest to the smallest droplets is more than 50:1.

It is for samples such as these, in which large size ranges of features are present, that images with a very large number of pixels are most essential. As an example, an image with a width of 500 pixels would realistically be able to include features up to about 100 pixels in width (20% of the size of the field of view), and down to about 20 pixels (smaller ones cannot be accurately measured). That is a size range of 5:1. But to accommodate 50:1 if the minimum limit for the small sizes remains at 20 pixels and the field of view must be five times the size of the largest (1000 pixel) features, an image dimension of 5000 pixels is required, corresponding to a camera of about 20 million total pixels. Only a few very high-resolution cameras (or a desktop scanner) can capture images of that size. It is the need to deal with both large and small features in the same image that is the most important factor behind the drive to use cameras with very high pixel counts for microstructural image analysis.

Furthermore, a 50:1 size range is not all that great. Human vision, with its 150 million light sensors, can (by the same reasoning process) satisfactorily deal with features that cover about a 1000:1 size range. In other words, we can see features that are millimeter in size and ones that are a meter in size at the same time. To see smaller features, down to 100 μm for example, we must move our eyes closer to the sample, and lose the ability to see large meter-size features. Conversely, to see a 100 meter football field we look from far off and cannot see centimeter size features. So humans are conditioned to expect to see features that cover a much larger range of sizes than digital cameras can handle.

There are a few practical solutions to the need to measure both large and small features in a specimen. One is to capture images at different magnifications, measure them to record information only on features within the appropriate size range, and then combine the data from the different magnifications. If that is done, it is important to weight the different data sets not according to the number of images taken, but to the area imaged at each magnification. That method will record the information on individual features, but does not include information on how the small features are spatially distributed with respect to the large ones. For that purpose, it is necessary to capture

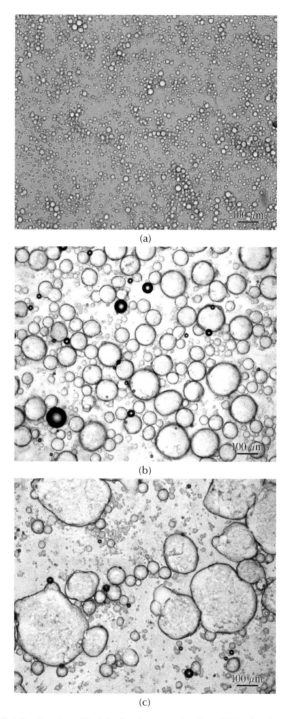

(a)

(b)

(c)

FIGURE 5.21 Fat droplets in milk: (a) after homogenization; (b) before homogenization; (c) before homogenization, long standing time. (Courtesy of Ken Baker, Ken Baker Associates)

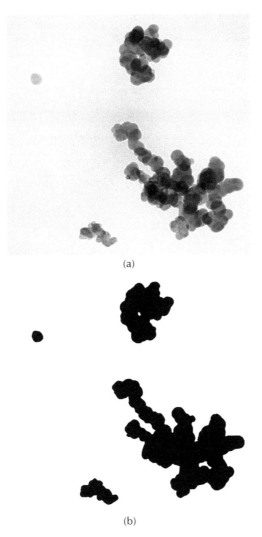

(a)

(b)

FIGURE 5.22 Using density to count particles in clusters: (a) original TEM image; (b) thresholded to delineate the clusters; (c) combination of images (a) and (b) to delineate the clusters with original grey scale values (labels give the integrated optical density and the estimated number of particles); (d) Beer's law calibration curve (the background point is set to the mean pixel brightness of the background around the clusters, corresponding to no attenuation).

multiple images at high magnification (which will show the small features) and then tile them together to produce a single large image in which the large features can be seen.

The images in Figure 5.21 raise another issue for feature measurement. Just what is the size of a droplet? The dark line around each droplet results because the index of refraction of light in water (the matrix) is not the same as in fat (the droplets). Consequently, each droplet acts as a tiny lens and light that passes through the edge of each droplet is directed away from the camera, producing the dark ring. The most

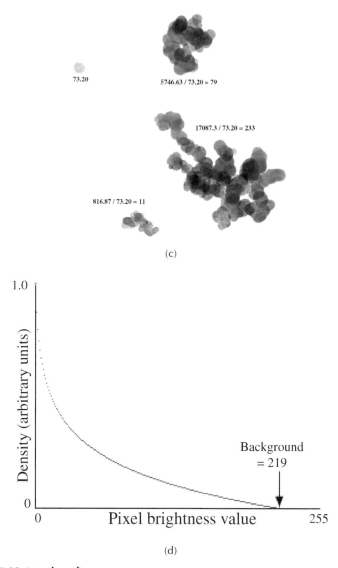

(c)

(d)

FIGURE 5.22 (continued)

convenient use of the dark outlines is to threshold them and fill the interiors of the droplets. That is correct for the case of fats in water. But for an emulsion of water in fat, the dark line would surround the droplet rather than being part of it. In that case, the procedure should be to threshold the dark line and process the image to keep the interior but not the line. So some knowledge of the nature of the sample and the physics of the generation of the image is needed to obtain accurate measurements.

Physics can be used to enable measurement in another situation that would otherwise be difficult, the case of a three-dimensional cluster of particles. In Figure 5.22

several clusters of particles, all about the same size, are seen (along with one image of a single particle). The projected area of each cluster can be measured, but is not meaningful. If this were a two-dimensional sample with clusters, the ratio of the area of the cluster to that of a single particle would provide a useful measure of the number of particles present in each cluster. But for a three-dimensional cluster this is not correct.

Physics describes the absorption of light (or any other radiation — the image is actually an electron micrograph) by Beer's law, an exponential attenuation of signal with the mass of material through which it passes. So measuring the integrated optical density (IOD) of each cluster and calculating the ratio of that value to the IOD of a single particle (preferably after measuring several to get a good average) will calculate the number of particles in each cluster.

Figure 5.22 shows the procedure. The original image is thresholded (based on the uniformity of the background) to produce a binary representation of the clusters. An opening was applied to remove isolated single-pixel noise. This binary image was then combined with the original to erase the background and leave each cluster with its original pixel grey scale values. A calibration curve was created using Beer's law of exponential absorption. The background point (zero optical density) was set to the mean brightness level of the background. The vertical density scale is arbitrary and does not require standards, since only ratios will be used.

The integrated optical density (IOD) of each cluster (and the single particle) is determined by converting each pixel value to the corresponding density and summing them for all pixels. The IOD thus represents the total absorption of radiation passing through the cluster, which is proportional to the total mass (and thus to the total volume) of material present. Dividing the IOD for each cluster by that for a single particle calculates the number of particles present in the cluster. Of course, this method only works if the cluster is thin enough that some signal penetrates completely through each point. If the attenuation is total so that a pixel becomes black, then more mass could be added without being detected.

Generally, this method works best with images having more than 8 bits of dynamic range. An optical density of 4 (the range that can be recorded by medical X-ray film) corresponds to attenuation values up to 99.99% of the signal, recording a signal of one part in 10,000. A camera and digitizer with 13 bits of dynamic range can capture a variation of one part in 8192, while 14 bits corresponds to one bit in 16,384. Cameras with such high bit depth are expensive, require elaborate cooling, and are generally only used in astronomy. A high-end, Peltier cooled microscope camera may have 10 or 12 bits of dynamic range (approximately one part in 1000 to one part in 4000).

BRIGHTNESS AND COLOR MEASUREMENTS

This foray into particle counting has brought us again to the measurement of brightness and/or color information from images. The example in Figure 5.10 showed counting of features based on their color signature. The example in Figure 5.22 showed the use of integrated optical density measurement for each feature (each cluster) to determine the number of particles contained within the feature. In general, density measurements are used to measure total mass. In the example of Figure 5.23, an image of frog blood cells shows dark nuclei and less dense cytoplasm. Because each cell has settled into a size and shape that only partly depends on its contents, and partly on its surroundings, measuring the area of each feature does not produce as useful a result as measuring its volume. This three-dimensional property can be obtained from the optical density measurement.

The cells in the original image touch each other. Rather than watershed segmentation, another procedure was used in this example. Each cell has a well defined, dark nucleus. These were thresholded and the image inverted so that the background could be skeletonized. The skeleton of the background is called the "skiz" and provides lines of separation that bisect the distance between features. Erasing the pixels along the skiz lines from the thresholded image of the entire cells separates them so they can be measured individually. The distribution plot of cell areas has a very different appearance from that of the cell integrated density values, which more accurately reflect the total cell volumes.

In many cases the density values, either calibrated as optical density or simple uncalibrated pixel brightness numbers, are used to measure profiles along lines across

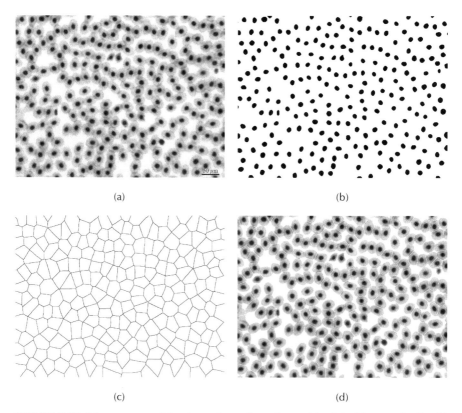

(a)

(b)

(c)

(d)

FIGURE 5.23 Measurement of density to determine volume: (a) original image of cells; (b) thresholded nuclei; (c) skeleton of the background (the skiz); (d) cells separated by erasing the skiz; (e) cell area distribution; (f) cell density distribution (arbitrary units).

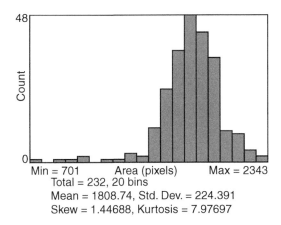

Min = 701   Area (pixels)   Max = 2343
Total = 232, 20 bins
Mean = 1808.74, Std. Dev. = 224.391
Skew = 1.44688, Kurtosis = 7.97697

(e)

Min = 114.8   Density   Max = 151.5
Total = 232, 20 bins
Mean = 134.689, Std. Dev. = 7.00476
Skew = –0.27632, Kurtosis = 2.59955

(f)

FIGURE 5.23 (continued)

the image. Figure 5.24 shows a typical example, measuring the density along a column in an electrophoresis separation of proteins. The values are averaged in the vertical direction at each horizontal point within the marked column, and the average intensity, usually converted to optical density, is plotted as shown. Analysis is then performed on the one-dimensional plot using peak finding software.

In many cases a plot of intensity is an effective way to extract information in a way that simplifies measurement. Growth rings in plants (Figure 5.25) provide a representative example.

Plots of color values can also be used in many cases to simplify measurement of dimensions. In the example of Figure 5.26, determining the thickness of the crust on the bread could be performed by a complicated procedure of thresholding and applying morphological operations to delineate a boundary, using a Boolean AND to apply a grid of lines, and then measuring the line lengths. It is much more efficient

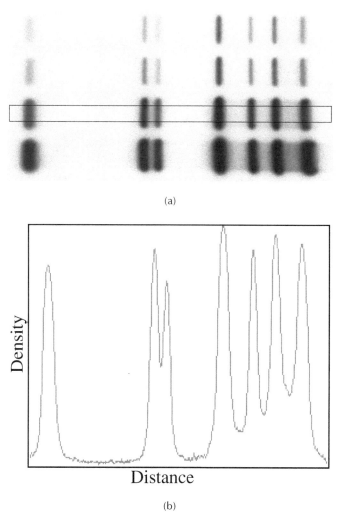

(a)

(b)

FIGURE 5.24 Plotting a density profile: (a) one column in a scanned image of protein bands separated by gel chromatography; (b) the resulting density plot.

to plot the average profile in a broad band perpendicular to the edge of the slice. There is a definite change in color associated with the crust, and it is (as usual) more effective to work with the hue, saturation and intensity values than with red, green, and blue. In the example, the hue is virtually unchanged, and the intensity changes only slightly at the boundary of the crust, reflecting primarily the presence of open cells in the bread. However, the saturation drops abruptly when the crust gives way to the interior bread structure, and provides a quick, effective and reproducible measurement of the dimension.

Calibrating the intensity vs. concentration for fluorescence images can be difficult, but in many cases can be avoided by instead using ratios of the intensity at two

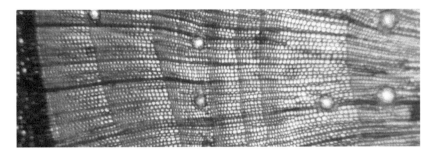

FIGURE 5.25 Growth rings in a fir tree can also be measured by plotting an intensity profile.

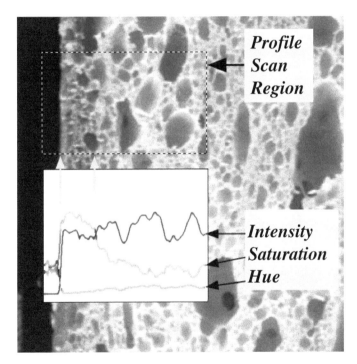

FIGURE 5.26 (See color insert following page 150.) Measuring the thickness of a surface layer (crust on bread) by plotting averaged pixel values. The saturation data are more useful than intensity or hue in this example.

different wavelengths of excitation, or two different emission wavelengths. Ratios of the intensity above and below the absorption edge energy of a dye, for example, provides a direct measure of the amount of the fluorescing material at each location. Figure 3.47 in Chapter 3 illustrates that process for a Fura stain that localizes calcium activity.

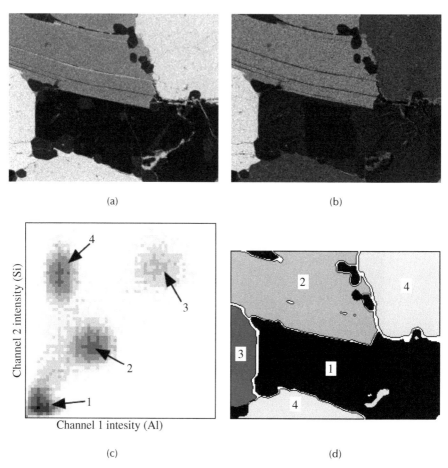

(a)   (b)

(c)   (d)

FIGURE 5.27 Co-localization plots: (a) Si X-ray map; (b) Al X-ray map; (c) co-localization plot. The four numbered clusters in the co-localization plot correspond to regions numbered on the diagram (d).

When two (or more) different dyes or stains are present, recording their emissions in different color channels is often used to produce color representations of the structure. Figure 2.15 in Chapter 2 shows a typical example. In a color image produced by recording different images in the red, green, and blue color channels, the co-localization of stains is indicated by color shifts or the presence of mixed colors such as magenta or yellow. A more quantitative approach to co-localization can be obtained by plotting the intensity values of the channels at each pixel in the image, as shown in Figure 5.27. The example in this case uses X-ray maps from an SEM, indicating the distribution of specific elements in the sample. Clusters in the plot represent pixels with various ratios of the two intensities, indicating the probable presence of structures or phases with unique chemical signatures.

LOCATION

The location of features can be specified in several ways. The absolute coordinates of the feature may be given in terms of distance from one corner of a slide or mounting grid, or as the position of the microscope stage X- and Y- drives. This is akin to using latitude and longitude as a way to describe position. For some purposes it is very useful, but in others it is more interesting to know the distance and direction from other features in the image. These in turn may either be similar features or other structures present such as cell walls or boundaries. Examples in Chapter 4 showed how the Euclidean map could be used to measure distance from boundaries, for instance.

Another issue that arises in specifying the position of a feature is just what point within the feature should be used for measurement. There are several possible choices, as shown in Figure 5.28. The centroid of a feature is the point at which it would balance if cut from uniform thickness cardboard, and this is one of the most widely used location parameters. But for a feature like a cell in which density varies from place to place (as discussed above, it typically measures local mass thickness), the density-weighted centroid may be a more logical choice. The geometric center can be specified in several ways, the most robust of which is the center of the circumscribed circle around the feature. Another possible location point is the ultimate eroded point from the Euclidean distance map; this is the center of the largest inscribed circle in the feature. For most features these points will not be in the same location and their measurement may lead to different interpretations of spatial distribution. If the features are symmetrical, such as spherical droplets, the points are all in the same place.

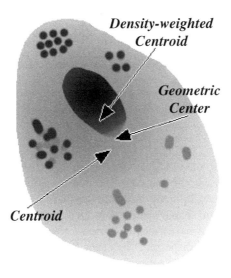

FIGURE 5.28 The location of a feature can be specified in several ways.

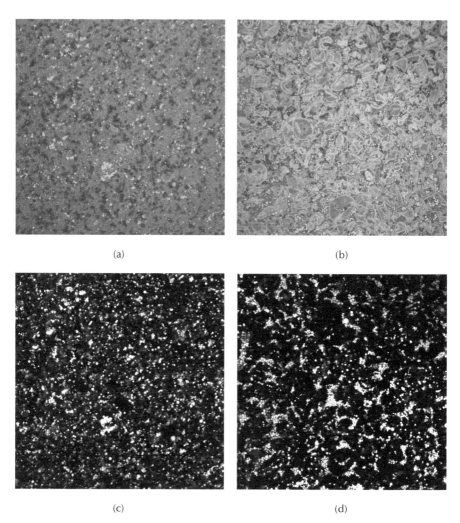

(a)   (b)

(c)   (d)

FIGURE 5.29 Measuring the spatial dispersion of fat in custard (courtesy of Anke M. Janssen, ATO B.V., Food Structure and Technology): (a, b) two custards (see Color Figures 3.56 and 4.7; see color insert following page 150); (c, d) thresholded red channels.

For some purposes, such as locating spots on electrophoresis gels, the absolute coordinates are important (in that case, the density-weighted centroid seems the most reasonable choice). In many more applications, it is neighbor distances that provide a way to measure spatial distributions. In the example shown in Figure 5.29, the two different custards have visually obvious differences in the uniformity with which the fat globules are dispersed. They have been processed by thresholding the red channel in a color image (because they were stained with Nile Red) and plotting the ultimate eroded points (the local peaks in the Euclidean distance map), which were then used as the locations of the features.

The nearest neighbor distances are calculated with the Pythagorean theorem from the absolute coordinates of the features in the image. A search is needed to locate the minimum value, but with modern computers this is very fast for images containing only hundreds or thousands of features. The key to using nearest neighbor distance as a measure of spatial distribution lies in the characteristics of a random distribution, in which each feature is independent and feels no force attraction or repulsion from the others (like sprinkling pepper onto a table). In that case, as shown in Figure 5.30, the nearest neighbor distances have a Poisson distribution. This is the hallmark of a random distribution. The Poisson distribution has very simple statistical properties, one of which is that the mean value can be calculated simply from the number of points. For a random distribution the mean nearest neighbor distance will be

$$Mean\ Nearest\ Neighbor\ Distance = \frac{0.5}{\sqrt{\dfrac{Number}{Area}}} \tag{5.3}$$

For the starch granules in potato tissue shown in Figure 5.30 the histogram has the shape of a Poisson distribution and a mean value of 35.085 (pixels). The calculated mean nearest neighbor distance for 447 features in an image that is 1600 × 1200 pixels in size is 32.76. These values are statistically indistinguishable for that number of features so we would conclude that the distribution of starch grains cannot be distinguished from random.

If the features are clustered together, the mean nearest neighbor distance is less than that calculated using the equation for the random distribution. Conversely if they are self-avoiding, the actual mean nearest neighbor distance is greater. The ratio of the actual mean nearest neighbor distance to the value calculated for a random spatial distribution is a useful measure of the tendency of the distribution toward clustering or self-avoidance. Since the distance values are compared only as ratios, the actual image calibrated magnification is not needed for this purpose.

Applying this procedure to the fat globules in the images of Figure 5.29 reports the following results:

Image	Actual Mean Nearest Neighbor Distance (in.)	Calculated Mean Distance for Random case (in.)
Figure 25(c)	0.1288	0.0985
Figure 25(d)	0.1254	0.1248

From these results it is evident that one of the two custards has a random distribution of fat globules, but in the other there is clustering present.

Conversely, Figure 4.41 of Chapter 4 showed a distribution of protein bodies and starch granules that were self-avoiding. There can be many reasons for clustering

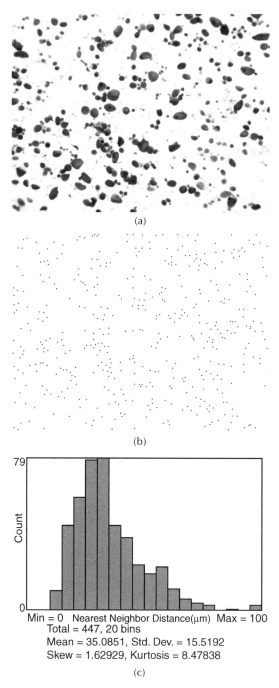

(a)

(b)

(c)

FIGURE 5.30 The spatial distribution of starch granules in potato: (a) magenta channel from the color image in Chapter 3, Figure 3.57 and Color Figure 3.57 (see color insert folllowing page 150); (b) ultimate eroded points for the starch granules; (c) distribution plot of measured nearest neighbor distances.

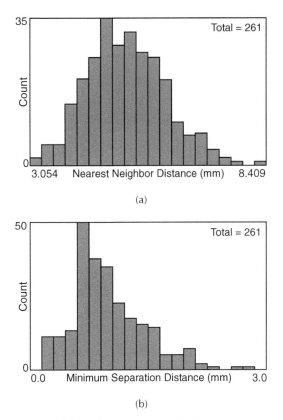

(a)

(b)

FIGURE 5.31 Plots of the distribution of nearest neighbor distance for the rice grains in Figure 5.14: (a) based on centroid-to-centroid distance; (b) based on the minimum edge-to-edge separation distance.

or self-avoidance, including chemical depletion, surface tension, electrostatic forces, etc. Finding that a non-random structure is present is only the first step in understanding the reasons for the structure to have formed. Figure 5.14 showed an image of rice grains dispersed for length measurement. The means of dispersal, a vibrating table, should produce mechanical interactions between the grains causing them to separate. Measuring the mean nearest neighbor distance produces a result of 5.204 mm, which is much greater than the value of 2.440 mm predicted for a random distribution. This confirms the (expected) self-avoiding nature of the dispersal.

In this image there is a significant difference between the nearest neighbor distance based on centroid-to-centroid spacing and the distance of closest approach or minimum separation between features. Figure 5.31 shows plots of the two different distances, and Figure 5.32 illustrates the meaning of the two parameters. In many structures, the identity of the feature that is nearest by one definition is not even the same as that which is closest by the other, which leads to some interesting but complex possibilities for performing statistical analysis on neighbor pairs. It is also interesting to observe that the nearest neighbor directions and the feature orientations in the rice image are not isotropic, as shown in Figure 5.33. In other words, there

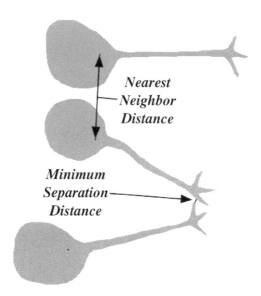

FIGURE 5.32 Diagram illustrating the nearest neighbor distance based on centroid-to-centroid spacing and edge-to-edge minimum separation, which may not be to the same neighbor.

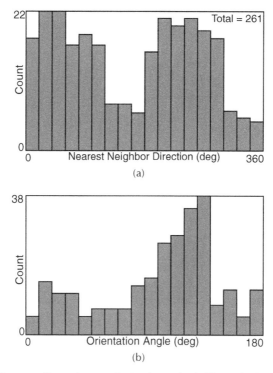

FIGURE 5.33 Plots revealing anisotropy in the rice grains in Figure 5.14: (a) nearest neighbor direction; (b) orientation angle of the grains.

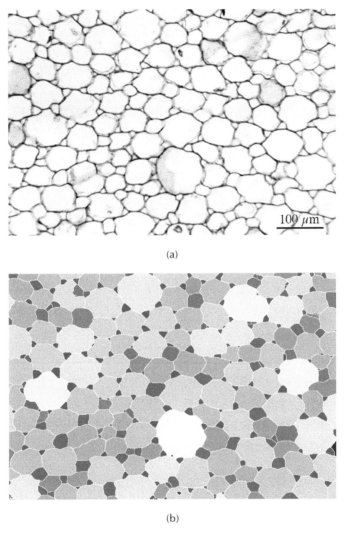

FIGURE 5.34 Section through corn plant tissue: (a) original image; (b) cells with grey scale values representing the number of touching neighbors; (c) plot of the number of sides (note the large number of three-sided cell sections); (d) regression plot of number of adjacent neighbors vs. cell size.

is a tendency toward alignment of the features, as might be expected from consideration of how the dispersal is accomplished.

Another type of neighbor relationship is the number of adjacent features, which is often the same as the number of sides on a polygonal feature. This arises most often in cell structures such as the corn plant section in Figure 5.34. In three dimensions the cells are polyhedra. Where the section plane cuts through the center of a cell, it produces a polygon with many sides and a large number of adjacent

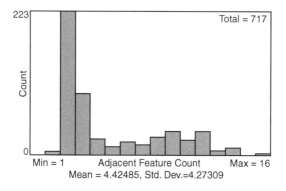

(c)

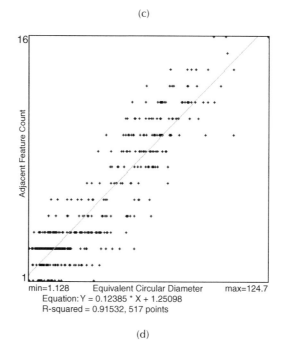

(d)

FIGURE 5.34 (continued)

neighbors. While the number of neighbors seen in the section is less than the number of touching neighbors in three dimensions, the analysis of number of neighbors still reveals some important characteristics of different types of structures. The maximum shown in the image is a cell with 16 neighbors. Where the location of the cut passes through a corner of the three-dimensional cell it produces a small section with a small number of sides. Note the very large number of three-sided sections, most of which occupy positions between the larger cells. As shown in the figure, there is a strong correlation between size and number of adjacent neighbors.

GRADIENTS

One major reason for interest in the location of features is to discover and characterize gradients in structure. Most natural materials, and many manufactured ones, are far from uniform, but instead have systematic and consistent variations in size, shape, density, etc., as a function of location. In many cases, the direction of the gradient is perpendicular to some surface or boundary. The complexity of structure and the natural variation in the size, shape, density, etc. of the individual features, can make it very difficult to detect visually the true nature of the gradient. Hence, computer measurement may be required.

Depending on the image magnification and the scale of the gradient, position may be measured by the Euclidean distance map or by simple X- or Y-coordinates. If the distance from some boundary or feature within the image is important, the EDM is the tool of choice. This was illustrated in Figures 4.42 to 4.44 in Chapter 4. For situations such as determining a gradient normal to a surface, if the image shows a section taken perpendicular to the surface, the Y-coordinate of a feature in the image may provide the required position information. In either case, plots of feature property vs. position are used to reveal and characterize the gradient.

As an example, Figure 5.35 repeats a diagram from Chapter 1 (Figure 1.13). Instead of counting the number of hits made by points in a grid to estimate the vertical gradient, as done there, procedures based on feature measurement will be used. A plot of area fraction vs. vertical position can be generated by counting the number of pixels covered by features at each vertical position (Figure 5.35c). However, that is not a feature-specific measurement. Visually, the nearest neighbor distance for each feature changes most strikingly from bottom to top of the image. Reducing each feature to its ultimate eroded point and measuring the nearest neighbor distance for each feature produces a graph (Figure 5.35d) that shows this gradient.

All of the features in the preceding example were identical in size and shape, it is only their distance from their neighbors that varies. More often the variation is in the size, shape and density parameters of the individual features. In the example in Figure 5.36, the size of each cell in plant tissue is measured. The procedure used was to threshold the image, skeletonize the binary result and then convert the skeleton to a 4-connected line that separates the cells. Measurement of the size (equivalent circular diameter) of each cell and plotting it against the horizontal position (of the centroid) produces the result shown in Figure 5.36(e). This plot is difficult to interpret, because of the scatter in the data. The band of small cells about 40% of the way across the width of the image is present, but not easy to describe. Interpretation of the data can be simplified by coloring each cell with a grey scale value that is set proportional to the size value (Figure 5.36d). A plot of the average pixel brightness value as a function of horizontal position averages all of the size information and shows the location of the band of small cells, as well as the overall complex trend of size with position.

Sometimes the color coding of features is not even required. In the example of Figure 5.37, an intensity plot on the original image suffices to show the structural gradient. The image is a cross-section of a bean. There is a radial variation in cell

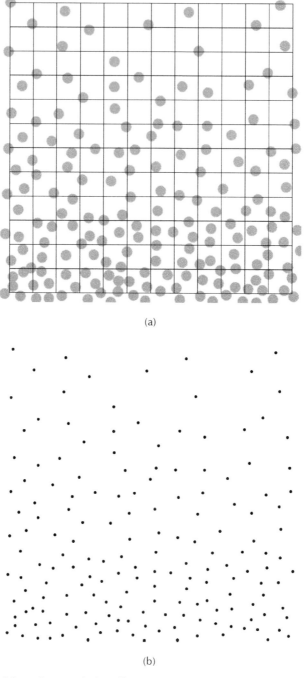

(a)

(b)

FIGURE 5.35 Measuring a vertical gradient: (a) original with superimposed grid; (b) ultimate eroded points; (c) plot of area fraction vs. position; (d) plot of nearest neighbor distance vs. position.

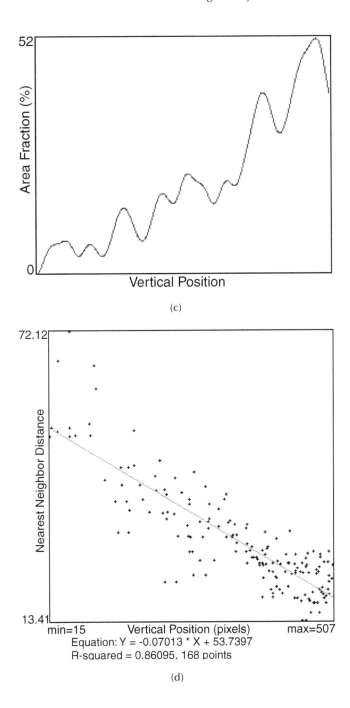

(c)

(d)

FIGURE 5.35 (continued)

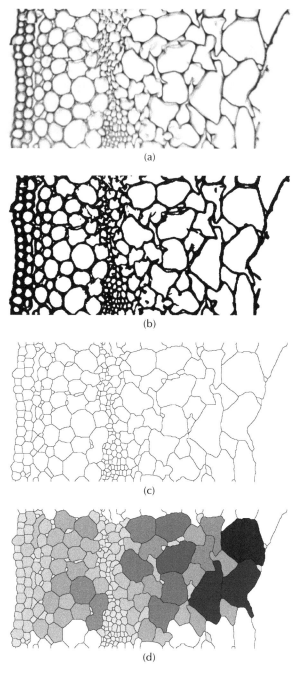

(a)

(b)

(c)

(d)

FIGURE 5.36 Measuring a complex gradient of size: (a) cross-section of plant tissue; (b) thresholded; (c) skeletonized; (d) cells colored with grey scale values that are proportional to size (equivalent circular diameter, but note that cells intersecting the edge of the image are not measured and hence not colored); (e) plot of individual cell values for equivalent diameter vs. horizontal position; (f) plot of average grey scale value vs. horizontal position.

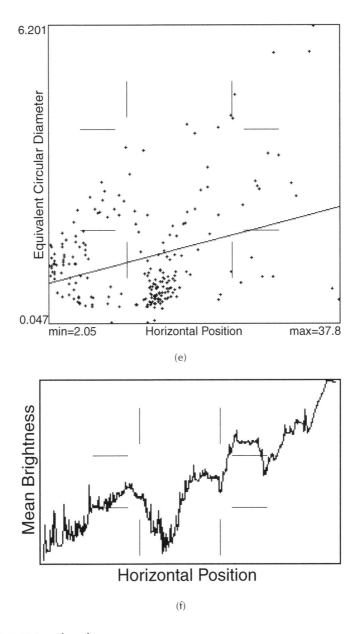

(e)

(f)

FIGURE 5.36 (continued)

size but this is visually perceived as a variation in brightness, and a plot of averaged brightness as a function of radius shows this structural variation.

When measurement of position is not simply an X, Y, or radial coordinate, the Euclidean distance map is useful for determining distance from a point of boundary. A cross section of natural material may be of arbitrary cross-sectional shape. Assigning each feature within the structure a value from the EDM measures its distance

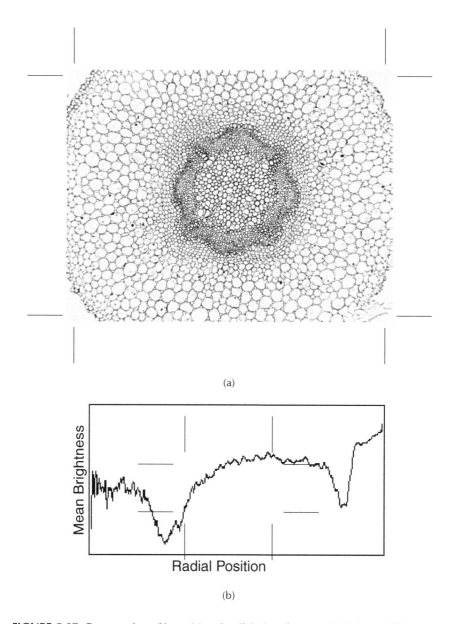

(a)

(b)

FIGURE 5.37 Cross-section of bean (a) and radial plot of averaged brightness (b).

from the exterior, as shown in Figure 5.38. This value can then be combined with any other measure of size, shape, etc. for the feature to allow characterization of the structure.

In addition to the problem of deciding on the direction of the gradient, it is often difficult to decide just which feature property varies most significantly along that direction. Figure 5.39a shows outlines of cells in fruit in a cross section image (the fruit exterior is at the top). In this case the direction of interest is depth from the

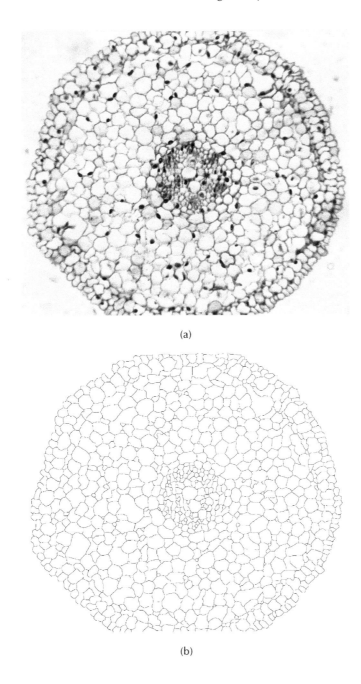

(a)

(b)

FIGURE 5.38 Using the EDM to measure location: (a) cross-section of a plant stem; (b) thresholded and skeletonized lines separating individual cells; (c) each cell labeled with the EDM value measuring distance from the exterior; (d) each cell labeled with a grey scale value proportional to cell size.

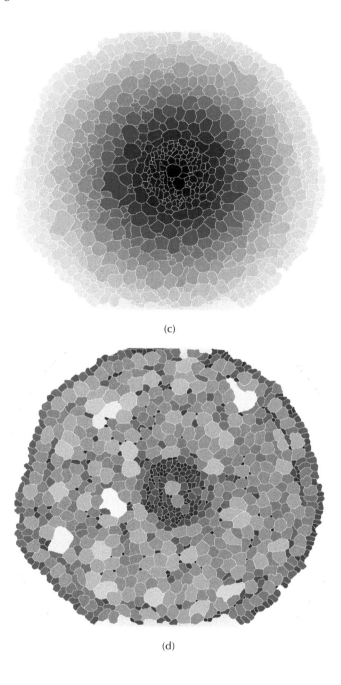

(c)

(d)

FIGURE 5.38 (continued)

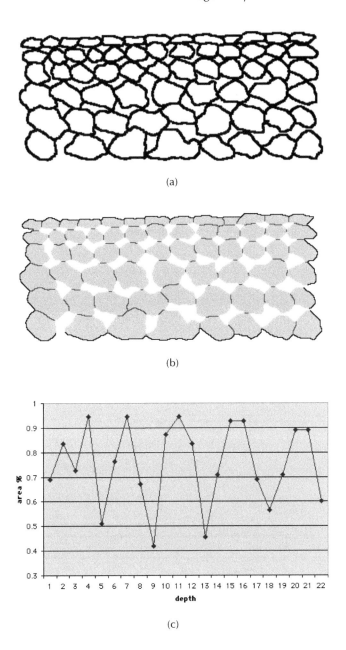

(a)

(b)

(c)

FIGURE 5.39 Cross-section of fruit showing a gradient in the structure of the cells and air spaces: (a) thresholded binary image; (b) cells with skeletonized boundaries coded to identify the portions adjacent to other cells or to air space; (c) plot of area fraction vs. depth; (d) plot of area fraction vs. cell layer, counting from the top; (e) plot of perimeter length (adjacent to another cell and not adjacent to a cell, hence adjacent to air space) vs. cell layer; (f) plot of fraction of cell perimeter that is adjacent to another cell vs. depth.

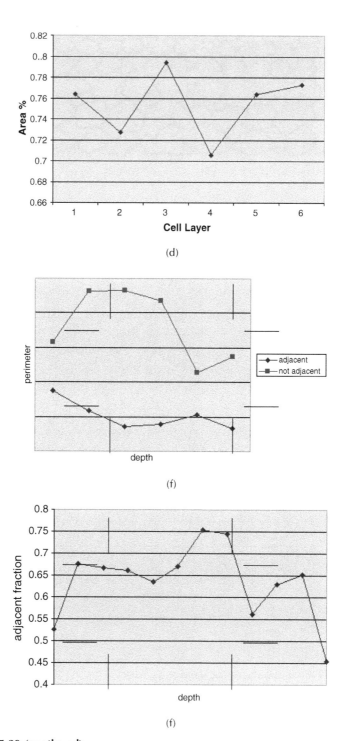

(d)

(f)

(f)

FIGURE 5.39 (continued)

surface of the fruit, but the choice of parameter is not at all obvious. This image was obtained by thresholding and then manually touching up an image of fruit, and seems typical of many of the pictures that appear in various journal articles and reports. Visually there is a gradient present, and the nature of that gradient probably correlates with properties including perception of crispness when biting into the fruit and perhaps to storage behavior. But what should (or can) be measured?

For a perfect fluid (liquid or gas) the deformation behavior is described simply by the viscosity and can be easily measured in a rheometer. For most real foods the situation is more complicated. The various components of the microstructure stretch with different moduli, fracture after different amounts of strain, interfere with each other during plastic flow, and generally produce small but important amounts of variation in the stress-strain relationship, which are often rate and temperature dependent as well. Measuring this behavior mechanically is challenging, and finding meaningful and concise ways to represent a complex set of data is important, but beyond the scope of this text.

The terminology in food science generally uses texture descriptors that are intended to correspond to the mouthfeel of the product during chewing. An example is the use of crispness for the magnitude of the fluctuations in stress during prolonged deformation at constant strain rate. Obviously this may result from many different factors, one of which is the breaking of structural units over time, either as an advancing fracture surface reaches them, or as they are stretched by different amounts until they reach their individual breaking stresses. Either of these effects might meaningfully be described as producing a crisp feel while biting into an apple. But a similar fluctuation would be observed in measuring the viscous behavior of a fluid containing a significant volume fraction of hard particles that interfere with each other, and that does not fit as well with the idea of crispness.

David Stanley has noted in reviewing a draft of this text that "the scientist trying to deal with definitions of texture and structure is often faced with the very difficult problem that extremely small changes in microstructure can cause huge changes in perceived texture. Our sensory apparatus is very sensitive, such that minute alterations in texture or flavour are perceived quite readily. With flavour, this may be a survival mechanism to help us avoid poisoning ourselves. In any case, it makes life hard for those looking to food structure as the basis for texture. It seems likely that these small changes in microstructure are a result of alterations in structural organization, i.e., the chemical and physical forces responsible for tenuous interconnections that are so easily broken and reformed during food processing operations. It is much easier to document and quantitate structure than structural interactions."

Allen Foegeding has also pointed out additional links between physical and sensory properties. For example, an important property of crisp and crunchy textures is sound. Even if we measure all of the properties associated with sound, appearance and texture, the brain can still perform some intricate and strange processing that defies simple statistical correlations. There is work going on, and more remaining to be done, concerning the link between mouth sensation and brain processing, as there is between mouth sensation and physical, mechanical and microstructural properties.

Certainly these observations are true, and it is not the intent here to oversimplify the problem, or to suggest that measurements of microstructural parameters by themselves will suffice to predict the mouthfeel of food. Summarizing a complex behavior by single numerical measurement may be convenient but it makes it more difficult to then find meaningful correspondences between the mechanical performance and the structural properties, which can also be measured and described by summary (usually statistical) values. In the case of the apple it might be the dimensions (length, thickness) of the cells or cell walls, or the contact of the walls with another cell, and perhaps the distributions of these values, while in the case of the particle-carrying fluid it might be the volume fraction, size distribution, and perhaps also the shape of the particles. These are all measurable with varying amounts of effort, and obtaining a rich set of measurement parameters for the microstructure makes it more feasible to use statistical methods such as stepwise regression or principal components analysis, or to train a neural net, to discover some important relationships between structure and performance.

Likewise the mechanical behavior needs to be characterized by more than a single parameter. In the case of crispness it might include not just the amplitude of the variations in the stress-strain curve, but also the fractal dimension of the curve and its derivative, and coefficients that describe the variation in those parameters with temperature and strain rate (to take into account the effect of these changes while chewing food). Further relating these mechanical parameters to the sensory responses of the people who chew the food, to isolate the various perceived effects, is a further challenge that appears much more difficult to quantify and depends to a far greater extent on the use of statistics to find trends within noisy data, and on careful definitions of words to establish a common and consistent basis for comparisons.

In measuring images to search for correlations between structure and behavior, there is an unfortunate tendency toward one of two extremes: a) measure everything, and hope that a statistical analysis program can find some correlation somewhere (although the likelihood that it can be meaningfully interpreted will be small); or b) bypass measurement, collect a set of archetypical images or drawings, and rely on humans to classify the structure as type 1, type 2, etc. (which is often not very reproducible and in any case still avoids the question of what are the meaningful aspects of structure). In this example skeletonization and the use of morphological and Boolean operations were used to identify the cells (grey) and label the periphery of the cells as either being adjacent to another cell, or to an air space (Figure 5.39b). The outer boundary (darkest grey line) is not included in the measurements shown.

Qualitative descriptions of these textures often mention the extent of the air spaces, and the size and shape of the cells, as being important factors. One of the most straightforward things to measure as a gradient is the area fraction (which as noted before is a measure of the volume fraction). Does the area fraction of the fruit occupied by cells (as opposed to air spaces) show a useful trend with depth? The plots in Figures 5.39(c) and 5.39(d) show the measurements, performed using a grid of points. In the first plot, the area fraction (which measures the volume fraction) is plotted as a function of depth. The values oscillate wildly because of the finite and relatively uniform size of cells in each layer, which causes the area fraction value to rise toward 100% in the center of a layer of cells and then drop precipitously in between. So the data were replotted to show area fraction as a function of the cell layer depth. But there is still no obvious or interpretable gradient. Apparently the qualitative description of a gradient of air space between cells is not based simply on area (or volume) fraction, at least for this example.

The plots in Figures 5.39(e) and 5.39(f) deal with the cell perimeters. The measured length of the perimeter lines in the image is proportional to the area of the cell surface. In the topmost plot the total length of these lines is shown, measured separately for the lines where cells are adjacent to each other and for those where a cell is adjacent to an air space. There is a hint of a trend in the latter, showing a decline in the amount of adjacent contact between cells, but it must be realized that such a trend, even if real, could occur either because of a drop in the total amount of cell wall, or in the fraction of cell walls in contact. Plotting instead the fraction of the cell walls that are in contact as a function of depth does not show a simple or interpretable trend (the drop-off in values at each end of the plot is a side effect of the presence of the surface and the finite extent of the image). Again, the qualitative description does not seem to match with the measurement.

In fact, people do not do a very good job of estimating things like the fraction of area covered by cells or the fraction of boundaries that are in contact, and it is likely that visual estimation of such parameters is strongly biased by other, more accessible properties. People are generally pretty good at recognizing changes in size, and some kinds of variation in shape. The plots in the Figure 5.40 show the variation in the size of the cells and the air spaces as a function of depth, and the aspect ratio (the ratio of the maximum to the minimum caliper dimension) as a function of depth. The two size plots show rather convincing gradients, although the plot for intercellular space vs. depth is not linear, but rises and then levels off. The aspect ratio plot shows that really it is only the very first layer at the surface that is significantly different from the others. Note that the measurements here are two-dimensional, and do not directly measure the three-dimensional structure of the cells, but because all of the sections are taken in the same orientation the measurements can be compared to one another.

Another characteristic of images that human vision responds to involves the spacing between features. As shown in Figure 5.41, measuring the nearest neighbor distances between the centroids of the cell sections shows a strong trend with depth, but in the absence of obvious changes in area fraction this is probably just dual information to the change in cell size with depth. For the air spaces, the ultimate eroded points were used as markers for the measurement of nearest neighbor distance. Because the shapes are not convex this seems to be a more meaningful choice.

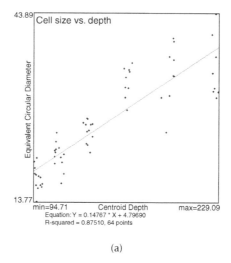

(a)

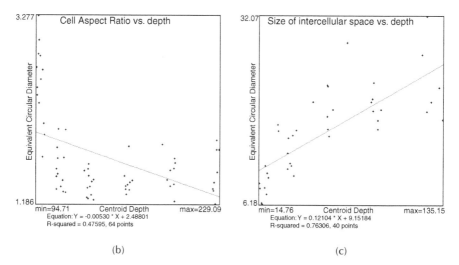

(b)   (c)

FIGURE 5.40 Plots of cell size (a) and shape (b) vs. depth, and plot of the size of intercellular air spaces (c) vs. depth.

The trend in the plot shows an increase in mean value, but it may be that the increase in variation (e.g., standard deviation) with depth is actually more meaningful.

As an indication of the difficulty of visually perceiving the important variable(s) involved in these textural gradients, Figure 5.42 shows another fruit cross-section (a different variety). There is enough complexity in the structure to make it appear visually similar to the first one. Seeing "through" the complexity to discern whatever underlying differences are present is very difficult.

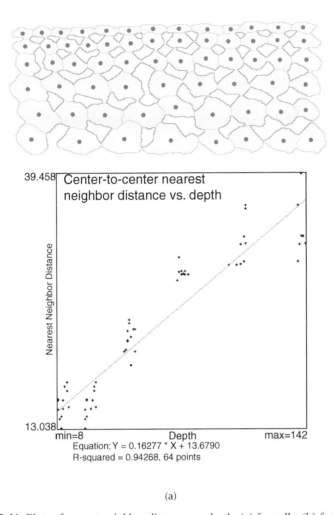

(a)

FIGURE 5.41 Plots of nearest neighbor distance vs. depth: (a) for cells; (b) for air spaces.

When measurements similar to those above are performed different results are observed. The trends of size and nearest neighbor spacing are even more pronounced than in the first sample, but there is only a weak shape variation with depth. Area fraction is still not correlated with depth, but the fraction of cell wall perimeter that is adjacent to another cell, rather than adjacent to air space, does show a trend in the first few layers near the surface.

These examples emphasize that many different kinds of information are available from image measurement. Generally, these fit into the categories of global measurements such as area fraction or total surface area, or feature-specific values such as

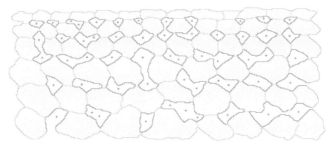

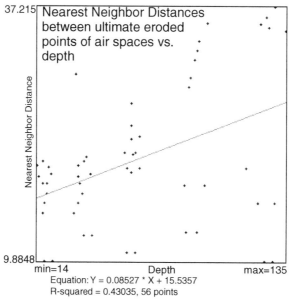

(b)

FIGURE 5.41 (continued)

the size, shape and position of each cell. Selecting which to measure, and relating the meaning of that measurement back to the structure represented in the image, requires thinking about the relationships between structural properties revealed in a cross-sectional image and the relevant mechanical, sensory or other properties of the fruit. Selection benefits from a careful examination of images to determine what key variations in structure are revealed in the images. These may involve gradients as a function of depth, which may be related to sensory differences between varietals or as a function of storage conditions.

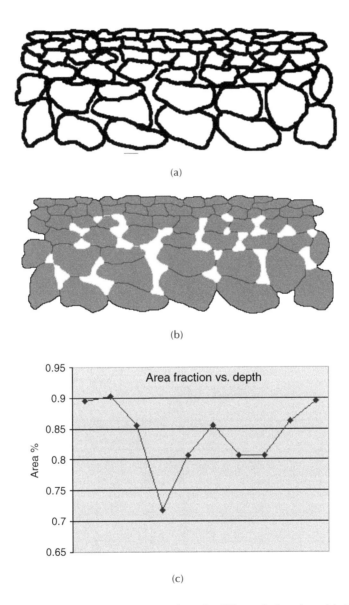

(a)

(b)

(c)

FIGURE 5.42 Measurement of the cross-section of a different fruit variety: (a) thresholded binary; (b) cells and air spaces with skeletonized boundaries identified as adjacent to another cell or adjacent to air space; (c) plot of area fraction vs. depth; (d) plot of fraction of the cell walls that are adjacent to another cell; (e) plot of cell size vs. depth; (f) plot of cell shape (aspect ratio) vs. depth; (g) plot of center-to-center cell nearest neighbor distance vs. depth.

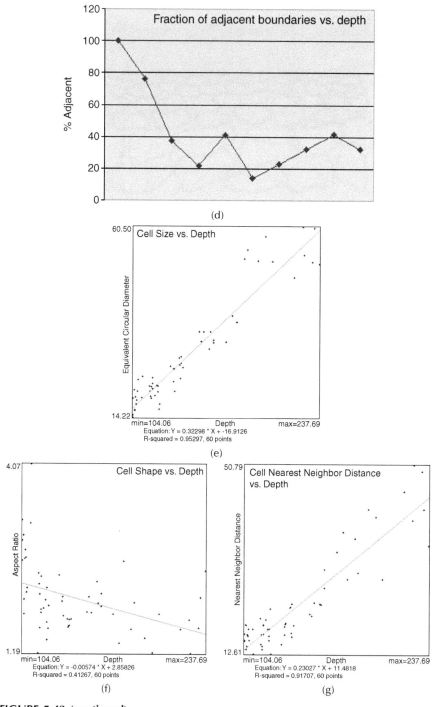

(d)

(e)

(f)

(g)

FIGURE 5.42 (continued)

SHAPE

In the preceding example, one of the parameters that varied with depth was the aspect ratio of the cells. That is one of many parameters that can be used to describe shape, in this case representing the ratio of the maximum caliper dimension to the minimum caliper dimension. That is not the only definition of aspect ratio that is used. Some software packages fit a smooth ellipse to the feature and use the aspect ratio of the ellipse. Others measure the longest dimension and then the projected width perpendicular to that direction. Each of these definitions produces different numeric values. So even for a relatively simple shape parameter with a familiar-sounding name, like aspect ratio, there can be several different numeric values obtained. Shape is one of the four categories (along with size, color or density, and position) that can be used to measure and describe features, but shape is not something that is easily translated into human judgment, experience or description.

There are very few common adjectives in human language that describe shape. Generally we use nouns, and say that something is "shaped like a ...," referring to some archetypical object for which we expect the other person to have the same mental image as ourselves. One of the few unambiguous shapes is a circle, and so the adjective round really means shaped like a circle. But while we can all agree on the shape of a circle, how can we put numbers on the extent to which something is shaped like (or departs from the shape of) a circle? Figure 5.43 illustrates two ways that an object can depart from circularity, one by elongating in one direction (become more like an ellipse), and the other by remaining equiaxed but having an uneven edge. There are more possibilities than that, of course — just consider n-sided regular polygons as approximations to a circle.

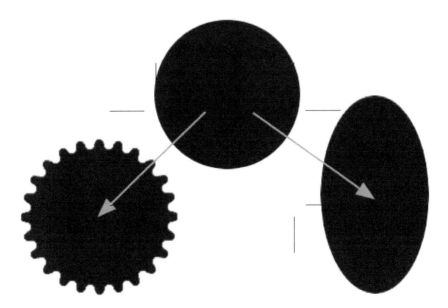

FIGURE 5.43 Two ways to vary from being like a circle.

TABLE 5.1
Derived Shape Parameters

Parameter name	Calculation
Formfactor	$\dfrac{4\pi \cdot Area}{Perimeter^2}$
Roundness	$\dfrac{4 \cdot Area}{\pi \cdot MaxDim^2}$
Aspect Ratio	$\dfrac{MaxDimension}{MinDimension}$
Elongation	$\dfrac{FiberLength}{FiberWidth}$
Curl	$\dfrac{Length}{FiberLength}$
Convexity	$\dfrac{ConvexPerim}{Perimeter}$
Solidity	$\dfrac{Area}{ConvexArea}$
Hole Fraction	$\dfrac{FilledArea - NetArea}{FilledArea}$
Radius Ratio	$\dfrac{InscribedDiam}{CircumscribedDiam}$

The departure from roundness that produces an uneven edge is often measured by a parameter called the formfactor, which is calculated from the area and perimeter, as summarized in Table 5.1. The departure that produces elongation is often measured by either the aspect ratio, or by a parameter usually called roundness. Actually, neither of these names (nor any of the others in Table 5.1) is universal. Other names like circularity, elongation, or compactness are used, and sometimes the equations are altered (e.g., inverted, or constants like π omitted), in various computer packages. The problem, of course, is that the names are arbitrary inventions for abstract arithmetic calculations.

Each of the formulas in Table 5.1 extracts some characteristic of shape, and each one is formally dimensionless so that, except for its effect on measurement precision, the size of the object does not matter. An almost unlimited number of these derived shape parameters can be constructed by combining the various size parameters so that the dimensions and units cancel out. Some of these derived parameters have been in use in a particular application field for a long time, and so have become familiar to a few people, but none of them corresponds very well to what people mean by shape.

Figure 5.44 shows a typical application in which a shape parameter (form factor in the example) is used. The powder sample contains features that vary widely in

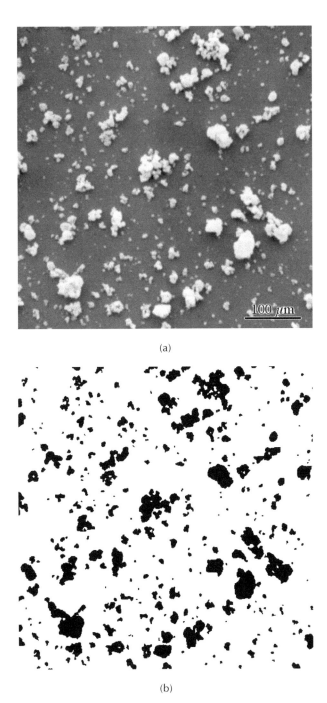

(a)

(b)

FIGURE 5.44 Shape and size of powder sample: (a) original SEM image; (b) thresholded; (c) size distribution (equivalent circular diameter); (d) shape distribution (formfactor); (e) regression plot of shape vs. size.

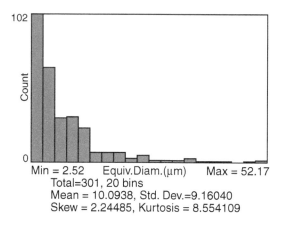

(c)

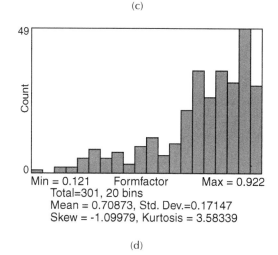

(d)

FIGURE 5.44 (continued)

size and shape. Histograms of the size and shape indicate that there are many small particles and many that are nearly round (a formfactor of 1.0 corresponds to a perfect circle). By plotting the size vs. the shape for each particle in the image, a statistically very significant trend is observed: the small particles are fairly round but the large ones are not. A common cause for that behavior (and the specific cause in this case) is agglomeration. The large particles are made up from many adhering small ones. There are many other types of shape change with size, in fact it is unusual for objects to vary over a wide size range without some accompanying shape variation.

Regression plots, such as the one in Figure 5.40, make the implicit assumption that the relationship between the two variables is linear. Nonlinear regression can also be performed, but still assumes some functional form for the relationship. A nonparametric approach to determining whether there is a statistically significant correlation between two variables is the Spearman approach, which plots the rank

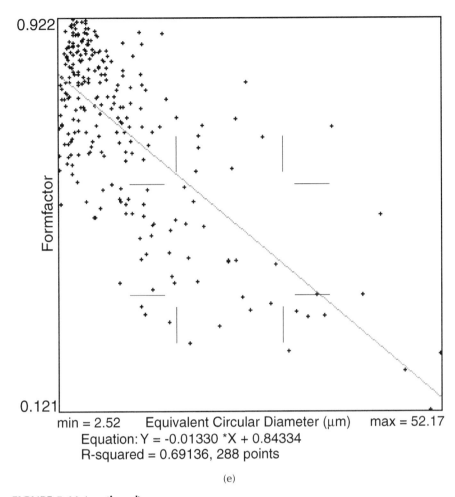

Equation: Y = -0.01330 *X + 0.84334
R-squared = 0.69136, 288 points

(e)

FIGURE 5.44 (continued)

order of the measured values rather than the values themselves. Of course, finding that there is a correlation present does not address the question of why.

One problem with the dimensionless derived shape parameters is that they are not very specific. Another is that they do not correspond in most cases to what humans mean when they say that features have similar or dissimilar shapes. As an example, all of the features in Figure 5.45 were drawn to have the same formfactor, yet to a human observer they are all extremely different shapes.

There are two other ways to describe some of the characteristics of shape that do seem to correspond to what people see. One of these deals with the topology of features, and those characteristics efficiently extracted by the feature skeleton. The second is the fractal dimension of the feature boundary. These are in some respects complementary, because the first one ignores the boundary of the feature to concentrate on the gross aspects of topological shape while the second one ignores the overall shape and concentrates on the roughness of the boundary.

FIGURE 5.45 A selection of shapes with the same numerical value of formfactor.

As shown in Chapter 4, the skeleton of each feature has end points with a single neighbor, and nodes points with three or four neighbors, which can easily be counted to determine the basic topology. Euler's relationship ties these together:

Number of loops =
Number of Segments – Number of Ends – Number of Nodes + 1 (5.4)

People extract these basic properties — especially the number of end points and the number of loops or holes in the feature — automatically and quickly by visual inspection. Computer measurement using the skeleton is nearly as fast.

In addition to the purely topological shape properties, the skeleton is used by itself and with the Euclidean distance map to determine other properties such as the mean length and width of segments, both the terminal and interior branches. Figures 4.41 and 4.42 in Chapter 4 showed examples of classification of features based on their topological parameters as measured by the skeleton.

Many natural objects have boundaries that are best described by fractal rather than Euclidean geometry. A formal definition of such a boundary is that the observed length of the perimeter increases with magnification so that a plot of the length of the total measured perimeter vs. the image resolution is linear on log-log axes. More practically, features are considered fractal if the apparent roughness stays the same as magnification is increased, at least over some finite range of scales (usually several orders of magnitude). This self-similar geometry arises from many physical processes

(a)

FIGURE 5.46 Fractal dimension measurement using the EDM: (a) the EDM of pixels inside and outside the classic Koch snowflake; (b) a table of the number of pixels as a function of their EDM value, and the constructed cumulative plot whose slope gives the fractal dimension.

such as fracture and agglomeration, but is certainly not universal. Features for which a single force such as surface tension or a bounding membrane dominates the shape will have a Euclidean boundary that is smooth and does not exhibit an increase in perimeter with magnification.

Examples of fractal surfaces and clusters were shown, along with procedures for measuring their dimension, in Chapter 3 (Figure 3.33) and Chapter 4 (Figure 4.42). Measurement of the boundary fractal dimension of features can be performed in several ways, which do not necessarily produce exact numerical agreement. The most robust method uses the Euclidean distance map to determine the number of pixels (both inside and outside the feature) as a function of their distance from the boundary. As shown in Figure 5.46(b), the slope of a log-log plot of the number of pixels vs. distance from the boundary gives the dimension, which varies from 1.0 (for a perfectly Euclidean feature) upwards. Typically rough boundaries often have dimensions in the range from 1.05 to 1.4; a few representative examples are shown in Figure 5.47.

There are other approaches to shape measurement as well. One that is rarely used but extremely powerful is harmonic analysis. The boundary of each feature is

Minowski Sausage Method				
EDM Value	Pixel Count	Cumulative	Log(R)	Log(A)
1	9330	9330	0.00000	3.96988
2	5956	15286	0.30103	4.18429
3	5198	20484	0.47712	4.31141
4	4052	24536	0.60206	4.38980
5	4448	28984	0.69897	4.46216
6	4008	32992	0.77815	4.51841
7	4070	37062	0.84510	4.56893
8	3416	40478	0.90309	4.60722
9	3580	44058	0.95424	4.64402
10	3620	47678	1.00000	4.67832
11	3640	51318	1.04139	4.71027
12	3796	55114	1.07918	4.74126
13	3096	58210	1.11394	4.76500
14	3260	61470	1.14613	4.78866
15	3074	64544	1.17609	4.80986
16	3114	67658	1.20412	4.83032
17	2978	70636	1.23045	4.84903
18	3082	73718	1.25527	4.86757
19	3072	76790	1.27875	4.88530
20	3000	79790	1.30103	4.90195
Fractal Dimension = 2 – slope =			1.28257	

(b)

FIGURE 5.46 (continued)

unrolled into a plot of direction vs. position, and Fourier analysis is performed on the plot (which of course repeats every time a circuit of the boundary is completed). The amplitude of the Fourier components provides a series of numbers that completely define the shape. In most cases only the first few dozen coefficients are significant (although for a truly fractal shape that would not be the case), and these numbers can be used in statistical analysis routines that search for discrimination between groups or underlying identities in complex mixtures.

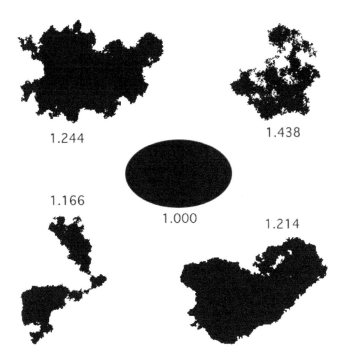

FIGURE 5.47 Several representative natural shapes (and one smooth ellipsoid) with the measured fractal dimensions for each.

The harmonic analysis method is used primarily in sedimentology, to classify particle shapes, but has been used occasionally on food products, for instance to identify different species of wheat kernels. The two principal reasons that it is not used more are: a) a significant amount of computation is required for each feature; and b) the list of numbers for each feature represent the shape in a mathematical sense, but are very difficult for a human to understand or decipher. To the author's knowledge, none of the image analysis programs that run on small computers include this function at the present time.

IDENTIFICATION

One goal of feature measurement is to make possible the automatic recognition or classification of objects. The cross-correlation method shown in Chapter 3 (Figures 3.41 and 3.42) can locate features in an image that match a specific target, but requires an image of the target that must be closely matched in shape, size, color and orientation. Consequently, while it can be a very powerful method for seeing through clutter and camouflage, it is not a generally preferred method for classification. Most techniques for feature identification start with the list of measurement parameter values rather than the image pixels.

There are many different techniques used. Most require prior training with data from representative examples of each class of objects, although there are methods that can find clusters of parameter values that indicate the presence of multiple classes, as well as methods that start with a few examples and continue to learn as more objects are encountered. There are many texts just on the various recognition and classification techniques, which are heavily statistical. Good introductions that cover a variety of methods include S.-T. Bow, *Pattern Recognition and Image Preprocessing* (Marcel Dekker, New York, 1992) and K. Fukunaga, *Introduction to Statistical Pattern Recognition* (Academic Press, Boston, 1990).

It is not the intent of this text to delve deeply into the relative merits of expert systems, fuzzy logic and neural nets, syntactical description, or kNN (nearest neighbor) methods. The following illustrations are basically implementations of a classic expert system but have some of the characteristics of fuzzy logic due to the use of histograms of feature parameters. Figure 5.48 shows a simple example, a collection of the letters A through E in different fonts, sizes and orientations. A set of four rules are shown connected in an expert system that identifies all of the letters.

This is a very sparse example of an expert system; most have hundreds or thousands of rules, and the task of the software is to find a logical connection path and efficient order of application that will result in paths from input values to output identification. But even this simple example indicates one of the limitations of the method: the difficulty of adding additional classes. If the letter "F" is added to the image, new rules are needed, some of the existing ones may be discarded, and the order of application may change completely.

Also, it is important to remember that the expert in an expert system is the human who generates the rules. These are typically based on experience with or measurement of many prototypical examples of each class (a training set). For instance, the rule shown for distinguishing the letters A and D is based on roundness. Figure 5.49 shows a histogram of the roundness values for the various A and D letters in the original image, showing that they are completely separated and indicating where a decision threshold can be placed between the two groups.

It is unusual to find a single parameter that completely separates the classes. One approach to dealing with this is to use statistical techniques such as stepwise regression or principal components analysis to find a combination of parameters that provide the separation. Instead of roundness, the horizontal axis for the histogram might be a combination of several measurement values, each multiplied by appropriate coefficients. Neural networks operate in a similar way, finding the weights (coefficients) that describe the importance of each input variable.

Another way to improve the separation of classes is to work with multiple parameters in a two-, three- or higher dimension space. For example, consider the task of separating cherries, lemons, apples, and grapefruit. Size clearly has a role, but by itself size is not sufficient for complete discrimination. Some lemons are larger than some apples, and some apples are larger than some grapefruit. Color is also a useful parameter. Cherries and apples are red, while lemons and grapefruit are yellow. But again there is overlap. Some grapefruit are pink colored and some apples (e.g., Golden Delicious) are yellow. Figure 5.50 shows a schematic approach to using both the redness of the color and the size of the fruit to separate the classes.

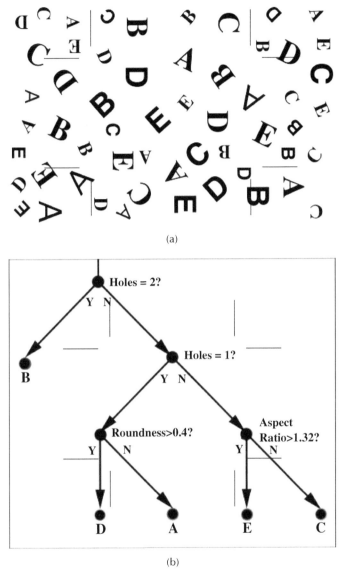

(a)

(b)

FIGURE 5.48 Expert system identification of features: (a) collection of letters A through E in various fonts, sizes and orientations; (b) rules that will successfully identify each feature.

Obviously, this method can be extended to more than two dimensions, although it becomes more difficult to represent the relationships graphically. Figure 5.51 shows a schematic example of a three-dimensional case in which classes are represented by boxes that correspond to ranges of size, shape and density parameters. Several parameters for each of these may be needed, especially for shape since as noted previously the various shape descriptors emphasize somewhat different aspects of feature shape.

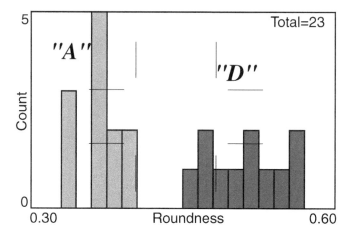

FIGURE 5.49 Histogram of roundness values for letters A and D.

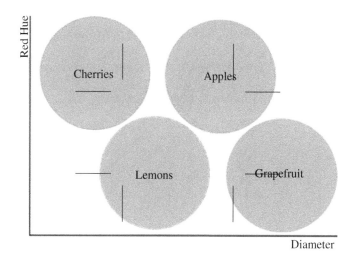

FIGURE 5.50 Schematic diagram for parameters that separate fruit into classes.

As an example of this type of classification, Figure 5.52 shows an image of mixed nuts from a can. There are known to be five types of nuts present. Many size and shape parameters can be measured, but which ones are useful? That question can be difficult for a human to answer. The plots in Figure 5.53 show that no single parameter can separate all of the classes, but also that each of the parameters shown has the ability to separate some of them. Determining which combination to use generally requires the use of statistical analysis programs that use regression to isolate the most important parameters.

Figure 5.54 shows a scatterplot of the measurements of two parameters, the inscribed radius (a measure of size) and the radius ratio (the ratio of the inscribed radius to the circumscribed radius, a measure of shape). In this plot the five types of nuts are

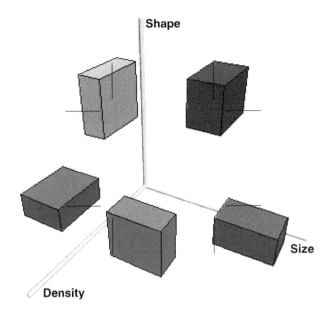

FIGURE 5.51 Separating classes in multiple dimensions.

FIGURE 5.52 (See color insert following page 150.) Image of nuts used for recognition.

completely separated. Figure 5.55 shows a very simple example of how numerical limits could be established for these two parameters that would identify each of the nuts.

There is always a concern about how representative the measurements on a small sample of objects used for training are of the larger population that will be classified.

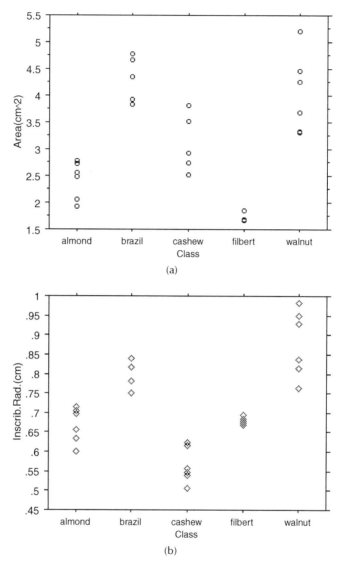

FIGURE 5.53 Plots of measured values sorted by type of nut: (a) area; (b) inscribed radius; (c) roundness; (d) formfactor.

Given a larger training population, better values for the discrimination thresholds would be determined, and it is possible that other parameters might be used as well to provide a more robust classification, but the principles shown do not change. Some systems are programmed to continue to learn, by adjusting the decision boundaries as more objects are encountered and the populations of each class become better defined.

The example in Figure 5.51 assumes that all objects examined will belong to one of the five classes of nuts for which the system is trained. A somewhat different

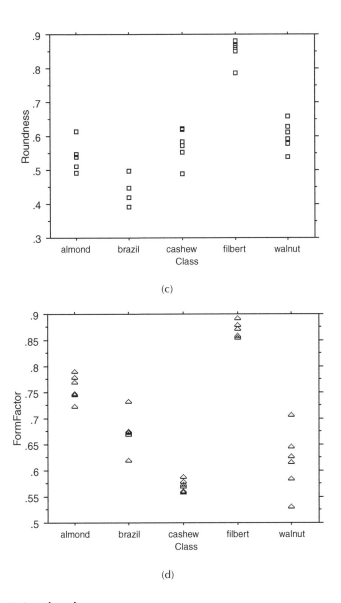

(c)

(d)

FIGURE 5.53 (continued)

approach using fuzzy logic with the same data is shown in Figure 5.56. The mean and standard deviation of the measured values from the training set define the ellipses that represent 95% probability limits for each class. A measured object that falls inside one of the ellipses would be identified as belonging to it. An object whose measured values fall outside the ellipses would be labeled as unclassified, but could be identified as most probably belonging to the nearest class, and the probability of that identification reported.

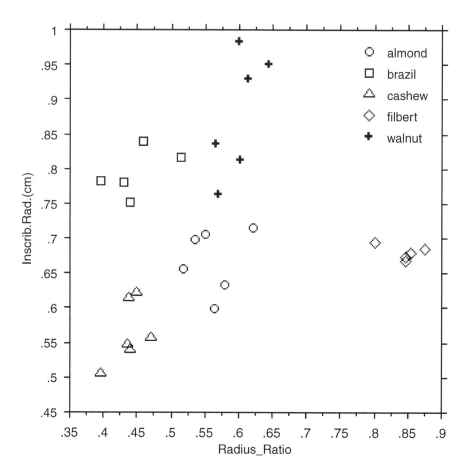

FIGURE 5.54 Two-parameter plot of inscribed radius and radius ratio (inscribed/circumscribed) separates the five nut classes.

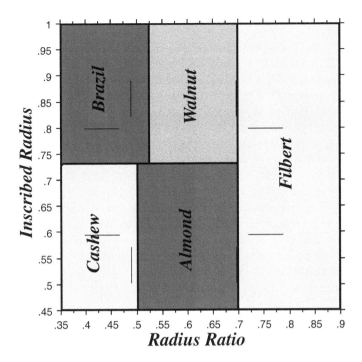

FIGURE 5.55 Parameter ranges that classify the five nut types.

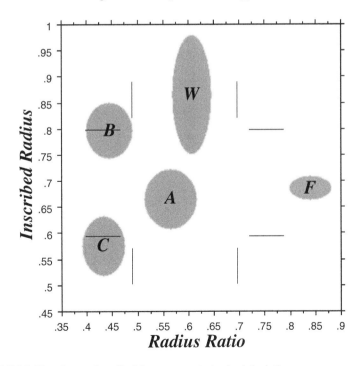

FIGURE 5.56 Nut classes described by mean and standard deviation.

CONCLUSIONS

In addition to the stereological measurement of structure, described in Chapter 1, there is often interest in the individual features present in an image. Counting the features present must take into account the effects of a finite image size, and deal with the features that are intersected by the edges. Additional edge corrections are needed when feature measurements are performed.

Feature measurements can be grouped into measures of size, shape, position (either absolute or relative to other features present) and color or density values based on the pixel values recorded. Calibration of the image dimensions and intensity values are usually based on measurement of known standards and depend on the stability and reproducibility of the system.

The variety of possible measures of size, shape, and position results in a large number of measurement parameters which software can measure. The algorithms used vary in accuracy, but a greater concern is the problem of deciding which parameters are useful in each application. That decision must rely on the user's independent knowledge of the specimens, their preparation, and the imaging techniques used.

Interpretation of measured data often makes use of descriptive statistics such as mean and standard deviation, histogram plots showing the distribution of measured values, and regression plots that relate one measurement to another. Comparisons of different populations generally require nonparametric procedures if the measured values are not normally distributed. Classification (feature identification) is also a statistical process, which usually depends on the measurement of representative training sets from each population.

Index

A

Acta Stereologica, 49
Alpha matrix values, 6, 7, 8
Aluminum foil, imaging of, 135, 137
Analog systems, limitations of, 65
Anisotropy, 22
  appearance of, 44
  of erosion/dilation, 220
  measurement of, 46
  in rice grains, 321
Apples, imaging of, 258
Area fraction, 11
  histogram of, 13
  plot of, 332, 336
  point grids used to measure, 243, 245
Area/volume, 24
Aspect ratio, 336, 342
Atomic force microscope (AFM), 48, 63
  color images with, 64–65
  problems of scale in, 121
  scan rates of, 67
  surface rendering of, 142, 144

B

Background image
  automatic correction for, 115, 119
  creating satisfactory, 112, 114, 115
  leveling contrast with, 113
  removal of bright, 115–116
  removal of structured, 115, 117
  selecting points for, 115, 118
Bayer filter pattern, for detectors, 58–59
Bean, cross-section of, 324, 328, 329
Beef, imaging of, 13, 52, 54–55, 136, 138,
    185–186
Beer's law, 310
Binary images, 277
  Boolean combinations, 236–240
  erosion and dilation, 213–219
  Euclidean distance map, 219–223
  fiber images, 260–262
  markers used to select features, 247–250
  region outlines as selection criteria, 252–254
  separating touching features, 223–235
  skeletonization, 254–259
  skeletons and feature shape, 262–263
  using grids, 240–246

Bit depth
  high, 54–55
  values for, 55–56
Blobs, 277
Blood vessels, imaging of, 241
Blur, types of, 123
Boolean AND, 312
Boolean logic operations, 176, 213, 236–240,
    250–252, 275, 277
Boundaries
  in contour method, 201, 204
  edges of features, 211
  manual marking of, 210
  for surface area, 22
Brain processing, mouth sensation and, 334–335
Branching, 39–40
Bread
  imaging of, 9, 90–91, 93–96, 210
  pores in, 36, 37, 204
Brightness
  abrupt changes in, 115
  calibrating, 288
  changes in, 84
  human perception of, 89
  measuring, 310–316
  and random noise, 99
  and surface roughness, 292
  and textural difference, 156
Brodatz textures, 157, 167
Bubbles, imaging of, 178, 198
Bug, imaging of, 147

C

Calcium activity, localization of, 179, 183
Calibration
  image, 287–292
  intensity, 288, 289
Cameras
  three-chip, 58 (*see also* Digital cameras)
  video, 57, 60
Candies, imaging of, 206, 249, 291
Candy bar, brightness histogram for, 14
Canny operator, 151, 153
Cavalieri method, 18, 20
CCD detector, 62
Cell membranes, imaging of, 197
Cell size, plots of, 336, 337
Cellulose fibers, imaging of, 164, 167

Chain-code perimeter, 294
Channel filtering, independent, 108
Channel merging, 84
Charge coupled device (CCD) cameras, 58
Cheese
    imaging of, 158
    SEM image of processed, 141
    thresholding fat in, 156, 158–159, 161
Chewing gum, imaging of, 100
Chocolate, imaging of surface of, 132, 144
CIE (Commission Internationale de l'Eclairage)
    color coordinates, 74, 77, 79
Circularity, departure from, 342–343
Closings, in erosion and dilation, 222–223
Clustering, measurement of tendency toward, 318
CLUT (color look up table), 142, 164, 195
Coffee particles, freeze-dried, 232
Collagen fibers, imaging of, 134, 166
Co-localization plots, 315
Color
    calibrating, 288
    false, 142
    human perception of, 78, 79
    measuring, 71, 310–316
Color adjustment, 71–77
    accuracy in, 72–74
    coordinate systems for, 74–77
    inverse matrix procedure, 73–76
Color channels, 81–86, *see also* RGB color
    channels
    and image combinations, 179, 181
    processing, 132
Color images, thresholding, 205–209
Color space coordinates, 77–81
Color values, 55–56
Comparisons
    K-S test, 301, 302
    non-parametric statistical, 301
Complementary metal oxide on silicon (CMOS)
    technology, 58, 62
Compression techniques, 67, 68
Computer graphics, manual marking, 210
Computers, image acquisition for, 51
Confection, imaging of, 283
Confocal microscopes
    color channel images from, 82–83
    color images of, 63–64
    fast scan rate of, 67
    point sampled intercepts in, 47
Constarch, imaging of, 199
Contour lines, 201, 202
Contour map, 201, 203
Contrast, improving local, 131–134
Contrast adjustment, automatic, 115, 120
Convex area, determination of, 294

Convolution methods, 145
Coordinate systems, for color space, 77–81
Corn plant, imaging of, 322–323
Cornstarch, imaging of, 113, 278, 279, 280
Cornstarch particles, size measurement on, 300
Counting procedure, 15, 18
    and density of particles, 280, 282
    disector, 285, 286–287
    edge correction in, 279
    for hidden features, 280–282
    particle, 310 (*see also* Particles)
    unbiased, 277–278
Crispness, 334
    and mechanical behavior, 335
    perception of, 334
Cross-correlation method, 350
    in image processing, 172–175
    and multiple images, 176
    shift, 176, 177
Crunchy textures, 334
CT, of human head, 180, 181
Cube, sections through, 9
Cubic particles, 6
Custard
    imaging of, 192, 193, 194, 317
    structural phase separation in, 222
Cycloid, 25
Cycloidal arcs, 25, 27
Cycloid grid, 23–24, 29, 30
Cylindrical surface, cycloidal, 29

D

Deconvolution, 122, 123, 124, 125, 127
Defocus blur, 123
Density measurements, 311–312
Density profile, plotting, 312–313
Descriptive statistics, 359
Design-based stereology, 11, 25
Detail, and erosion and dilation, 215–218
Detectors
    CCD, 62
    CMOS, 62
    for digital cameras, 57–58
    performance of, 54
Difference-of-Gaussians (DoG) filter, 137, 139,
    140
Digital cameras, 57
    CCD, 62, 63
    CMOS, 62, 63
    color filters used in, 72
    color images acquired by, 58–60
    dynamic range of, 63
    size range for, 60–61
Dilation, 213–219

Dinner mint, imaging of, 281
Directionality, 164–169
Disector count, 285, 286–287
Disector logic, extended, 39–40
Disector measurement, 11, 34, 36, 37, 38
Distortions, acceptable, 121
Distributions
    log-normal curve, 297–299
    normal, 297
    size, 5–11, 297–301
Droplets, measuring, 306–39
Dye-sub printers, 80
Dynamic range
    of scanners, 53
    with scanning microscope, 63–65

E

Edge-correction methods, 279
    adjusted count in, 304
    and large features, 305, 306
    older procedure for, 303
    and small features, 306
Edge filtering, 153, 154
Edge-finding, 147–155
Edge location, 151
Edges, aliasing of, 179, 182
Eggplant, reading color value for, 56
Egg shell membrane, imaging of, 164, 165
Eight-connected convention, 277
Electron microscopes, 2, 121, *see also* Scanning
    electron microscope
Electrophoresis gels, locating spots on, 317
Emulsions, 22
Epoxy resin, bubbles in, 149
Equiaxed cells, 44
Erosion and dilation
    and EDM, 220–221
    and rank neighborhood operations, 213–219
Erosions, and particle size, 226–227
Euclidean distance map (EDM), 219–223
    eroded points on, 317
    fractal dimension measurement using, 348–349
    location point on, 316
    to measure location, 328, 330–331
    for measuring gradients, 324
    measuring with, 263–275
    and perimeter measurement, 295–296
Euler characteristic, 40
Expected values, 4
Experiment design, 31–34
Extended focus, 121, 123
Extrusion, direction of, 25

F

Fast Fourier Transform (FFT), 104, 106, 110
Fat, spatial dispersion of, 317
Fat crystals, imaging of, 271
Fat droplets, imaging of, 10, 187, 192–193, 194,
    306, 307
Fat globules, in mayonnaise, 225
Feature-AND, 247, 248, 252
Feature boundaries, as test probes, 252, 253
Feature counting, unbiased, 278–279
Features
    dilation of, 214
    edge-touching, 296
    EDM measurement of, 264–265
    erosion of, 214
    in image enhancement, 169–175
    measuring color of, 290, 291
    in opaque volume, 283
Features measurement
    brightness, 310–316
    calibration, 287–292
    color, 310–316
    comparisons, 301–302
    counting, 277–287
    edge correction, 303, 359
    gradients, 324–341
    identification, 350–358
    location, 316–323
    shape, 342–350
    size distributions, 297–301
    size measurement, 292–297
Feature tracking, by image combination, 179, 184
Feret's diameter, maximum, 293, 294
Fibers, 28
    cellulose, 164, 167
    in image enhancement, 164, 167
    imaging of, 164, 167, 230, 234–235, 259,
        260–261
    measurement of, 294–296
    muscle, 244
Fields of view, systematic random location of, 33
File formats, image, 67, 70, 71
Filing system, designing, 129
Fixed pattern noise, 98
Flavour, 334, *See also* Mouthfeel
Fluid, deformation behavior of, 334
Fluorescence images, 313
Foamed product
    imaging, 4
    network in, 38
    stereological measurement of, 4–6
Food, texture of, 156, *see also* Textures
Food samples, chemical staining of, 192
Food structures, *see also* Structure

preferred orientation of, 43
volume imaging of, 1
Four-connected relationship, 277
Fourier power spectrum, 169, 170
Fourier space, averaging in, 169, 170
Fourier-space procedures, for image enhancement, 130
Fourier transform, 102, 110–111, 122
fast, 104, 105, 106, 110
inverse, 169, 170
for periodic noise, 127
Fractal compression, 68
Fractal geometry, 159, 161, 163
Fracture, examining surfaces produced by, 279, 280
Frequency-space representation, of image, 169–170, 171
Fruit
cross-section of, 340–341
imaging of, 332–333
measurement of, 2–4
and separation of classes, 351–353
Fruit fly, imaging of, 123
Fura stain, 179, 183, 314

G

Gas, deformation behavior of, 334
Gaussian smoothing filters, 101, 103, 104, 114
Gels, 22, 28
Geometric probability, 2
Geometry, fractal, 159, 161, 163
Gradients
averaged brightness and, 324, 328, 329
and cell perimeters, 336
computer measurement of, 324
in cross-section of fruit, 329, 332–333, 334
measuring complex, 327–328
measuring vertical, 325–326
textural, 337, 340–341
Grey scale intensity, 144
Grid procedure, 15–17, 19
Grids
computer-generated, 277
for measuring layer thickness, 240
Growth rings, in plants, 312–313, 314
Guard frame method, for edge correction, 303, 304

H

Harmonic analysis, 348–350
Herringbone textile pattern, 168, 169
Hi-pass filters, 140
Histogram modification, 130
Histograms, 127
of area fraction, 13
automatic settings, 189–197
brightness, 14
changing peak heights of, 189
for color space, 205, 207
defined, 12
of EDM image, 266, 269
equalization, 97
and image contrast, 86, 87, 88
low contrast on, 87, 90–91, 93–96
with threshold values, 190–191
Histogram stretching techniques, 100
Holes, counting and measuring, 280, 281, 284
HSI color channels, 85, 86
HSI (hue-saturation-intensity) space, 68, 77, 78, 193
Hue
for bread crust, 313
and color calibration, 290–291
Hue channels, 68, 84, 85
Hue colors, and fiber angles, 168

I

ICC (International Color Consortium) curves, 79
Ice cream, imaging of, 124, 246
Ice crystals
in ice cream, 297–299
imaging of, 246
Identification of features, 359
expert system, 351, 352
separation of classes in, 351–354
techniques for, 350
Illumination, nonuniform, 111–120
Image acquisition
color adjustment in, 71–77
color channels for, 81–86
color space coordinates in, 77–81
corrections in, 127
digital cameras for, 57–63
and distortion, 121–127
file formats for, 67–71, 127
film-and-scanner approach, 57
nonuniform illumination in, 111–120
optimum image contrast in, 86–97
removing noise in, 98–111
with scanners, 51–57
from scanning microscopes, 63–67
Image analysis, computer-based, 277
Image combinations, 175–184
Image compression, 67
Image contrast, optimum, 86–97
Image enhancement
with automatic thresholding, 197–200
in color image thresholding, 205–209
directionality in, 164–169

edge-finding, 147–155
finding features in, 169–175
foods available for, 130
global procedures for, 130
image combinations, 175–184
image sharpening, 135–141
improving local contrast, 131–134
with local variance equalization, 132, 133
manual marking, 209–210, 211
purpose of, 129
with rank-based filters, 145–147
texture in, 156–163
thresholding, 184–189
Image magnification, for surface area, 24
Image plane
length of line in, 22
with surfaces, 21
Image ratios, and combining images, 179
Images
manual measurement of, 209–210, 211
storage of, 71
Image sharpening
directional derivative for, 135, 136, 137
with hi-pass filters, 141
and Laplacian operator, 137, 138, 139
and overall contrast, 137–139
top-hat filter for, 145
Inkjet printers, 80
Integrated optical density (IOD), 310
Intensity
for bread crust, 313
and color calibration, 290–291
Intercept lengths, 44, 45
Interlace shift, correcting, 176, 178
Interpretation
descriptive statistics for, 359
stereological, 4
Intestine, stained, 193, 194
Isolate gel networks, imaging of, 159, 162, 163, 174, 175
Isotropic grid, 25
Isotropic sampling, 24, 25
IUR (isotropic, uniform, random) probes, 31, 49

J

Journal of Microscopy, 49
JPEG (Joint Photographers Expert Group) compression, 67, 68, 69

K

Koch snowflake, 348
Kolmogorov-Smirnov (K-S) nonparametric test, 301, 302

Kruskal-Wallis test, 301
Kurtosis, 297

L

L-a-b coordinate system, 68, 77, 78
Laplacian operator, 137, 138, 139
Laser printers, 80
Latex spheres, imaging of, 172, 173
Length
in 3D, 29
of S-shaped fiber, 294, 296
Light microscope images, distortion of, 121
Light microscopes, 2
Line length fraction, 12, 15
Lines
in stereological analysis, 28–31
and surface area, 21
Liquids
bubbles in, 149
deformation behavior of, 334
imaging of, 197
Liver tissue, TEM image of, 156, 160
Loaf of bread, size of voids in, 18, *see also* Bread
Local operations, for image enhancement, 130
Local variance equalization, 132, 133
Location of features, 316–323
Lossless image formats, 68, 71
Low pass smoothing filter, 106
Luminance, plotting, 290
LZW (Lempel-Ziv-Welch) compression, 68, 70–71

M

Mann-Whitney procedure, 301
Maximum caliper dimension, 293, 294
Maximum likelihood filter, 153, 155
Mayonnaise
fat droplets in, 10, 225
imaging of, 192, 194, 225
oil droplets in, 61–62
Mean intercept length, 44, 46
Mean reciprocal intercept value, 243
Meat
processed, 28
veins in, 28
Medical imaging, 177, 180–181
Metal alloy, histogram of, 188
Metric properties, 34, 35
Milk, fat droplets in, 307
Minimum caliper dimension, 294
Morphological operations, 213, 214, 275
Motion blur, 127
defined, 123
removal of, 125

Mouthfeel
    predicting, 335
    and surface area per unit volume, 201
    and texture descriptors, 334–335
MRI, of human head, 180
Muscle
    imaging of, 171
    pork, 242, 255
    sections through, 45

N

Nearest neighbor distance, 318
    with centroid-to-centroid spacing, 321
    distribution of, 320
    and gradient measurement, 324, 325
    plots of, 336, 338, 339
Needle-like cells, 44
Neighborhood operations
    for image enhancement, 130
    sharpening methods, 145
Neighborhood size, and noise reduction, 104–105,
        107
Neighbor relationships, with adjacent features,
        322–323
Networks
    extended, 40
    topological properties of, 38
New stereology, 11
Noise
    defined, 98
    random speckle, 99, 108–109
    removal of extraneous, 214, 215, 218
    removal of periodic, 110
    single-pixel, 310
    tradeoff with sharpness, 125
Noise reduction
    on grey scale image, 218, 219
    and image enhancement, 129
    methods, 100–102
    and neighborhood size, 104–105, 107
Nuclepore filters, 173
Number per unit area, 277
Nuts
    imaging of, 353–354
    and separation of classes, 354–358

O

Oil droplets, size range of, 60–61
Openings, in erosion and dilation, 222–223
Optical density, 310, 311
Optics, and nonuniform illumination, 112

P

Packaging, image of blisters on, 171, 172
Particles
    measuring clusters of, 309
    plotting shape of, 345
Particle size. measurement of, 226–227
Peanut, imaging of, 17, 118, 268
Peltier cooling, 63
Perimeter measurements, 294
PET, of human head, 180
Pixels, 57
    Boolean combinations of, 236
    brightness values of, 89
    in camera detector, 59
    EDM of, 272, 273
    on histogram, 12–13, 197
    for low magnification imaging, 60
    and noise reduction, 106–109
    non-square, 121
    and perimeter length, 294
    and thresholding, 184
Pizza
    imaging of, 208
    quality control inspection of, 74, 75
Plane probes, isotropic orientation of, 29
Plants
    growth rings of, 312–313, 314
    imaging of, 327–328
Plant stem, imaging of, 330–331
Plate-like cells, 44
Points, in stereological analysis, 28–31, 29, 31
Point-sampled intercept technique, 11, 45
Poisson distribution, 318
Polynomial method, 115, 118, 120
Pores
    imaging of, 9
    network of, 28
Pork, imaging of, 150, 242, 244, 255
Potato
    histogram for, 318
    imaging of, 193, 257, 319
Powder sample, shape and size of, 343–345
Printers, types of, 79–80
Printing technology, 80–81
Projections, 277
Proteins, electrophoresis separation of, 312, 313
PSD file format, 71
Pythagorean theorem, 318

Q

Quadruple points, in bubble structure, 28

R

Randomization, achieving, 31
Random speckle noise, 129
Rank-based procedures, 145–147
Raster scanning, 63
Retinex processing, 132–134
RGB color channels, 86
    in chemical staining, 193
    and noise reduction, 108, 110
RGB color coordinates, 77, 209
RGB values, 72–74, 81
Rice grains
    imaging of, 1, 114, 321
    measurement of, 296, 297
    nearest neighbor distance for, 320
Rose plot, 44, 46

S

Saddle point, 40
Salmon, imaging of, 266
Sample preparation, for image acquisition, 51
Saturation
    for bread crust, 313
    and color calibration, 290–291
Saturation channels, 84, 85
Scanners
    and bit depth, 55–56
    dedicated film, 56–57
    detectors in, 54
    for digital imaging, 57
    dynamic range of, 53
    types of, 51–52
Scanning electron microscope (SEM), 48, 63
    color images with, 65
    and random noise, 98–99
    scan speed of, 65–66, 67
    spatial averaging with, 100
Scanning microscopes, 67
    raster scanning of, 63–64
    scan speed of, 65
Scientific imaging, 68
Second-order stereology, 11
Section measurements, 277
Segmentation, 185, *see also* Thresholding
    of angular particles, 232–233
    of touching features containing holes, 227–231
    watershed, 223, 224, 225, 227, 234–235,
        250–251
Selection
    based on multiple criteria, 238
    Boolean, 239
Self-avoidance, measurement of tendency toward,
    318
Shannon method, 190–191

Shape
    derived parameters for, 343
    and formfactor, 347
    fractal dimensions of, 350
    harmonic analysis of, 348–350
    of particles, 345
    for powder sample, 343–345
    roundness, 342–343
    and separation of classes, 354–356
    statistical analysis programs for, 353
    and topology of features, 346
Shortening
    imaging of, 271
    linear structure of, 28
Shot noise, 109, 110, 127, 218
Signal-to-noise ratio, 98
Size, measurement of, 292–297
Size classes, 5–6, 8, 10
Size distributions
    for abnormal shapes, 300–301
    alpha values in conversion of, 7
    unfolding, 5–11
Skeletonization, 213, 254–259
Skew, 297
Skin, scanned probe image of, 143
"Skiz," 311
Smoothing filter, 109
Sobel magnitude filter, 150–151, 152
Sobel orientation filter, 164, 166, 167, 195, 196
Soy protein isolate particles, imaging of, 106
Spatial resolution
    defined, 52
    with scanning microscope, 63–65
Spearman approach, 345–346
Spectrophotometer, 71
Sphere
    calculation of size distribution for, 8–11
    sectioning, 5, 9
Sphere unfolding, 11
Split-and-merge strategy, 200–201
Starch granules
    histogram for, 318
    imaging of, 1, 17, 193, 279, 280, 319
Statistical analysis, limitations of, 1
Stereological events, 277
Stereological relationships, 4, 6, 48–49
Stereology
    defined, 2
    experiment design in, 31–34
    lines and points in, 28–31
    need for, 1
    surface area in, 21–27
    topological properties in, 34–41
    unfolding size distributions in, 5–11
    volume fraction in, 11–20

Stress-strain curve, 335
Structure
    definition of, 334
    topological properties of, 34, 35
Student's t-test, 301
Surface
    enhancement, 142–144
    fractal dimension for, 159, 163, 348
Surface area, measurement of, 21–27
Surface displays, 142, 143
Surface gloss, measurement of, 291–292
Surface layer, measuring thickness of, 314
Systematic random sampling, 31–34

T

Taffy
    imaging of, 134
    network structure in, 39
TEM image, of liver tissue, 156, 160
Tessellation, skeletonization of, 255
Texture-detecting filters, 156, 158–159
Textures
    definitions of, 334
    gradients of, 337, 340–341
    in image enhancement, 156–163
    qualitative descriptions of, 335–336
    and surface area per unit volume, 201
    visually different, 156, 157
Thickness
    calculation of, 42, 43
    and intersection lengths of line probes, 42, 43
Thick slice sample, 29, 30
Three-dimensional structures
    measurements of, 277
    unfolding size distributions for, 5–11
Thresholding, 211
    automatic, 197–200
    automatic settings, 189–197
    in Boolean selection, 239
    and brightness values, 185, 187
    with contour method, 201, 202
    defined, 184
    directional, 167–169
    in features measurement, 317
    histogram of, 186–187, 188
    interactive setting, 207, 208
    red channel for, 192
    setting for, 185–186
    and size uniformity, 200
    smooth boundary, 198
    split-and-merge strategy, 200–201
    thinning after, 195
    of three populations, 195–197
    with two levels, 187

Thresholding, color image, 205–209
TIFF (tagged image file format), 70, 71
Top-hat filter, 145–147, 169, 173, 211
Topological properties, stereological measurement
        of, 34–41
Touching, definition of, 277
Transfer function, 89, 93–94, 97
Transmission imaging, 48
Transmission microscopy, 2, 29
Trapezoidal distortion, correction for, 121, 122
Triple lines, in bubble structure, 28
Tristimulus camera, 290
Tristimulus coefficients, 75, 76
Trussell method, 190, 191
T-statistic, 191
Tubular structure
    determining length of, 34
    measuring, 30

U

Ultimate eroded points (UEP), 227, 228
Unbiased stereology, 25
Unfolding, 47
    defined, 5
    size distributions, 5–11
    sphere, 11
    technique, 8, 10
Unit volume, calculating length per, 42

V

Vertical sections method, 25, 26, 29
Video cameras, 57, 60
Viewing modes, 283–285
Vignetting, and nonuniform illumination, 112
Vision, human
    of depth, 336
    edge finding, 147–148
    limitations of, 142
    perception in, 141
    size range for, 60, 306
    and variations in brightness, 211
Volume
    Cavalieri measurement of, 18, 20
    measurement of density to determine, 311–312
Volume fraction, 11–20, 39, 185
    of bread pores, 38
    calculation of, 11
    counting procedure for, 12, 14–16
    gradients in, 18
    grid procedure for, 17
    methodology for measuring, 12
    point grids used to measure, 243, 245
Volume probe, 34

W

Watershed segmentation, 223, 224, 225, 227, 234–235, 250–251
Wavelet compression, 68
Whey protein, imaging of, 159, 162, 163, 174, 175
Whipped foam product. imaging of, 198
Whipped food product, histogram of, 88

Wiener deconvolution, 125, 126
Wilcoxon procedure, 301

Z

Z-bands, in muscle tissue, 171
Zone system, 89

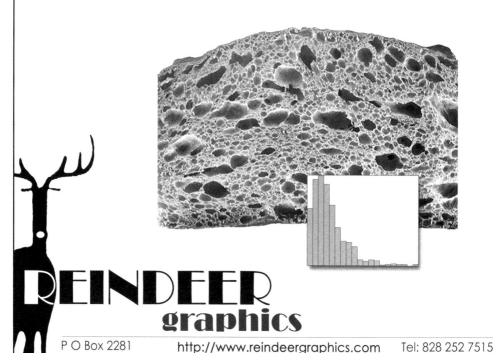

T - #0350 - 071024 - C15 - 234/156/18 - PB - 9780367393595 - Gloss Lamination